The Geography of Soils

W9-AOD-360

The Geography of Soils

formation, distribution, and management

DONALD STEILA

East Carolina University

Prentice-Hall, Inc., Englewood Cliffs, New Jersey

Library of Congress Cataloging in Publication Data

STEILA, DONALD, (date)
  The geography of soils.

  Bibliography: p.
  Includes index.
  1. Soil geography. 2. Soil science. I. Title.
S591.S834   631.4   76-230
ISBN 0-13-351734-9

10 9 8 7 6 5 4 3 2 1

Printed in the United States of America

Prentice-Hall International, Inc., *London*
Prentice-Hall of Australia Pty. Limited, *Sydney*
Prentice-Hall of Canada, Ltd., *Toronto*
Prentice-Hall of India Private Limited, *New Delhi*
Prentice-Hall of Japan, Inc., *Tokyo*
Prentice-Hall of Southeast Asia Pte. Ltd., *Singapore*

To my wife Becky
and
my daughter Stephanie Cristina

Contents

Preface

Throughout the United States there has been a rapidly developing interest in environmental quality and the recognition of the important role which soil serves as one of man's primary resources. As a consequence many Liberal Arts students are now including an introductory soil science course in their curriculums. In the process of teaching such a course—*Geography of Soils*—to students with limited training in the natural sciences, it became apparent that a need existed for a textbook written at a level comprehensible to the nonspecialist, but one that was not superficial in scope. It is my sincere hope that I have approached meeting these objectives.

Many individuals have assisted in the development of this book. Special recognition should be given to Dr. Jack Blok (Director, ECU Cartographic Laboratory), who personally directed the production of all graphic renderings, and to the East Carolina University Staff Cartographers Robert Corbo and Steven Moore. Professors Ronald Swager and Vernon Smith provided assistance when schedules were rushed. Andy Goodwin read the chapters and made valuable contributions to each one. Georgia Arend was a research aide and typed the final manuscript. Fellow colleagues and graduate students at the University of Georgia, University of Arizona, and East Carolina University provided suggestions for improvement. Most important in the production of this manuscript, however, has been my family, who gave constant encouragement and were understanding when research and writing demanded much of my time.

The Geography of Soils

Introduction

> Next to the pursuit of peace, the really greatest challenge
> to the human family is the race between food supply and
> population increase. That race tonight is being lost.[1]

It long has been recognized that the world's production of food
is lagging behind the population's rapid increase. The time required
for the earth's number of inhabitants to double decreases with the
passing of each year. It took at least a million years for the human
population to reach its size in 1650. Only 200 years later, in 1850,
this number had doubled. Presently, it is doubling in only 35 years.
If recent trends continue a population of more than 6 billion by the
year 2000 is estimated even if population growth is slowed.

This burgeoning population will require additional living space
and a considerable increase in food production to meet its projected
needs. Reporting in *The World Food Problem*, the President's Science
Advisory Committee stated: "The scale, severity, and duration of the
world food problem are so great that a massive, long-range, innovative
effort unprecedented in human history will be required to master it."[2]
The greatest pressures upon the land are expected to occur in the
developing countries of the world. By 1985 India will be

---

[1] President Lyndon B. Johnson, State of the Union Message, January 10,
1967.

[2] President's Science Advisory Committee, *The World Food Problem*, vol.
1, *Report of the Panel on the World Food Supply* (Washington, D.C.: U.S. Govt.
Ptg. Office).

faced with an 88 to 108 percent increase in food needs due to increased population; Brazil's food needs are estimated to increase by 91 to 104 percent; and Pakistan's by 118 to 146 percent. In each of these cases the lower percentage value assumes a 30 percent reduction in fertility while the higher value is derived from current population trends. These percentages do not reflect increased food requirements to improve dietary deficiencies, but simply the quantity needed to sustain future population growth at current levels of consumption.

It seems safe to estimate that by the year 2000 the world's population will require at least twice the present food supplies. The ability to meet this demand ultimately rests with the soils of the continental landmasses. Next to air and water, soil is probably the most fundamental earth resource that man has.

Soil serves as an anchorage for plants and as their nutrient reservoir. Both organic and inorganic in composition, soil is the loose surface material of the earth in which many complex biological, chemical, and physical processes take place. Stated most simply, soil is made up of four components: (1) organic matter, (2) inorganic material, (3) water, and (4) air. Each of these components varies from site to site in both character and amount. This variation is due to the interaction of climate and vegetation with inorganic materials on different geomorphic surfaces. When these interactions occur within relatively homogeneous areas, the soils of the region have certain characteristics in common. However, smaller scale environmental differences within a region can additionally impart a unique character to individual soil units.

The Origin and Significance of the Soil's Inorganic Constituents

1

MINERALS AND PARENT MATERIAL

An ideal loam soil is comprised of approximately 45 percent mineral matter. This mineral portion is the basic framework of the soil, serving as both an anchorage and a nutrient reservoir for plants. As such, the origin and characteristics of minerals are of prime concern to the *pedologist* (soil scientist).

A mineral is generally defined as a naturally occurring element or compound formed by inorganic processes and having a crystalline structure. Most of the earth's minerals are aggregated, or clustered, into types of rock. For example, granite is a common rock type that is composed of many interlocking crystalline minerals, including quartz, micas, and orthoclase and plagioclase feldspars. While each mineral has unique characteristics, every one had its origin in exactly the same way.

As the earth evolved from a molten sphere into a solid globe, cooling took place. All that cooling actually involved was a reduction in the activity of the *ions*[1] that comprised the *magma* (molten rock). With decreased activity, the ions responded to their electrical attractions and became bonded together in a fixed position, producing solid crystalline minerals. The composition of the magma and the resultant minerals was exactly the same, but in the solid state

[1] An ion is an atom, group of atoms, or compound that is electrically charged as a result of the loss or gain of electrons.

3

the ions were arranged according to a definite pattern, or crystalline structure.

All of the naturally occurring elements (special combinations of the protons, neutrons, and electrons) now recognized on earth were present in the molten mass from which the earth formed. Eight elements dominated the composition of the magma, however, and they now make up over 98.5 percent of the earth's crust by weight (Figure 1.1).

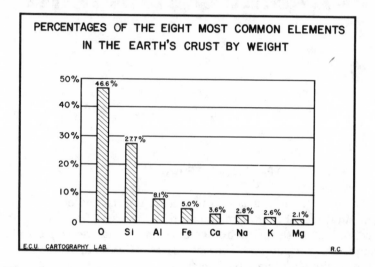

Figure 1.1 Percentages of the common elements in the earth's crust by weight—O = oxygen, Si = silicon, Al = aluminum, Fe = iron, Ca = calcium, Na = sodium, K = potassium, and Mg = magnesium.

Various combinations of the earth's elements produced a wide variety of minerals. The more important original and secondary types are outlined in Table 1. Original minerals have been produced dominantly in primate rock types—that is, *igneous*; however, the secondary minerals are primarily the result of weathering (including the formation of sedimentary rocks). As they occur in their aggregate structure, as rock, original and secondary minerals form the bulk of the earth's outer crust. When minerals, through subsequent weathering, are released from their bond with adjacent mineral crystals they may be available to supply the major bulk of soil material and the nutrient needs of plants.

Parent Material and Its Transformation

The processes leading to the development of a true soil require a considerable length of time. Numerous alterations must be per-

Table 1   Selected Original and Secondary Minerals

Original Minerals

Name	Formula*
Quartz	SiO_2
Microcline	$KAlSi_3O_8$
Orthoclase	$KAlSi_3O_8$
Na-plagioclase	$NaAlSi_3O_8$
Ca-plagioclase	$CaAlSi_3O_8$
Muscovite	$KAl_3Si_3O_{10}(OH)_2$
Biotite	$KAl(Mg \cdot Fe)_3Si_3O_{10}(OH)_2$
Horneblende	$Ca_2Al_2Mg_2Fe_3,Si_6O(OH)_2$
Augite	$Ca_2(Al \cdot Fe)_4(Mg \cdot Fe)_4Si_6O_{24}$

Secondary Minerals

Name	Formula
Calcite	$CaCO_3$
Dolomite	$CaMg(CO3)_2$
Gypsum	$CaSO_4 \cdot 2H_2O$
Apatite	$Ca_5(PO_4)_3 \cdot (Cl,F)$
Limonite	$Fe_2O_3 \cdot 3H_2O$
Hematite	Fe_2O_3
Gibbsite	$Al_2O_3 \cdot 3H_2O$
Clay minerals	Al-silicates

*For those unfamiliar with chemical formulae for minerals the following example may be helpful: Quartz has the formula SiO_2. This indicates a chemical bond between 1 silicon atom (Si) and 2 oxygen atoms (O_2).

formed on the surface layer of the earth's crust before a soil capable of supporting plant life is developed. Such changes normally involve the *disintegration* and *decomposition* of exposed rock material. Disintegration signifies a reduction in size of the original material, while decomposition refers to the chemical alteration of minerals. Collectively, the processes of disintegration and decomposition are called *weathering*.

Since a soil's characteristics can be strongly dependent upon the materials in the underlying collection of minerals, the original mineral complex from which a soil is formed (and is still forming) is called the soil's *parent material*.

There are two basic groups of inorganic parent material. One is known as *sedentary* (residual) and the other *transported*. A sedentary parent material is one that is native to the site. Suppose, for example, that a granite outcrop is weathering in place, without any significant removal of material. The soil that is formed will be composed of the residual products of the parent material and will, therefore, be considered sedentary (Figure 1.2). On the other hand, many soils occur on inorganic material that originated somewhere else. A river such as the Mississippi, when overflowing its banks, may deposit sediments that have been transported hundreds, even thousands, of miles. Such parent material is not sedentary (Figure 1.2). Besides running water, gravity (down slope movement), glacial ice, waves and offshore currents, and wind may carry inorganic material to a foreign site. Collectively, this type of parent material is said to be transported.

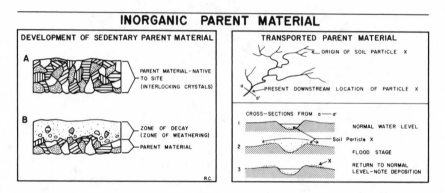

Figure 1.2 Examples of the difference between sedentary and transported parent material.

Weathering Processes

When affected by weathering, solid rock is broken into smaller rock fragments and ultimately into the individual minerals that originally comprised the rock structure. As size is altered minerals also can be modified or completely changed chemically. These processes that either release or alter particles and make them available for soil development fall into two broad groups—*mechanical* and *chemical weathering*.

We know that mechanical processes involve a physical reduction in the size of rocks and a separating out of the minerals. Several factors are significant:

Unloading. This is a process in which the removal of overlying rocks or sediment reduces pressure on the freshly exposed rock, permitting it to expand, thereby producing cracks and fissures.

Temperature variation. Since each mineral expands and contracts at a different rate when heated or cooled, rocks in environments that experience wide diurnal (daily) temperature ranges usually develop stresses between the surface mineral bonds—eventually weakening and separating individual crystals. However, this aspect of weathering is significant only when moisture is present, even in minute amounts.

The intense heat of forest and brush fires can raise surface rock temperatures dramatically within a few minutes, causing a steep temperature gradient in the upper rock layer and contributing to severe rock rupture. Frost shattering occurs in regions where there are periodic freezes. Water freezing into cracks, crevices, and pore spaces exerts tremendous pressure as it expands, and this can pry apart or even shatter massive rocks.

Plants and animals. Plant roots penetrating into cracks and crevices exert a prying effect on rock material as they develop, and animals burrowing or dislodging earth fragments also aid the process of physical disintegration. However, the influence of plants and animals is minor compared to the other mechanical weathering processes.

Chemical weathering, on the other hand, covers a group of processes in which minerals are altered in composition and reduced in size, transforming the original material into something different. The main processes involved are hydrolysis, hydration, oxidation, and carbonation. Seldom do they operate individually, but rather, are interrelated. If this were not the case, weathering would be an extremely slow process.

Hydrolysis. This refers to the process in which dissociated H^+ (hydrogen) and OH^- (hydroxyl) ions of water react with many rock-forming minerals. The effect of water in altering minerals is without doubt the dominant chemical weathering activity. Following is an example of the alteration of the mineral orthoclase. When precipitation (H_2O) comes into contact with the mineral orthoclase [$K(AlSi_3O_8)$], the hydrogen ion from the water may disrupt the mineral's crystal structure, producing an aluminosilicic acid ($HAlSi_3O_8$) and a hydroxide (KOH). The aluminosilicic acid, which is unstable, undergoes further change, which may result in the formation of clay minerals through recrystallization.

Hydration. When water combines chemically with other molecules, hydration occurs. Although this process may change the mineral structure, it often affects only the surfaces and edges of mineral grains without modifying their overall structure. An example of mineral conversion is the change of anhydrite ($CaSO_4$) in the presence of water (H_2O) to gypsum ($CaSO_4 \cdot 2H_2O$).

Oxidation. In this process oxygen combines with compounds in the rocks to form oxides. As oxidation takes place, the original material rots and is weakened. Iron, titanium, manganese, copper, and phosphorous are the dominant elements of rocks and soils that experience oxidation and reduction. Oxidation of iron, for example, produces a corrosive substance commonly known as rust. Normally the oxidation process is slow. However, with increased temperatures (such as in tropical regions) it is much more rapid.

Carbonation. Precipitation (H_2O) plus carbon dioxide (CO_2) unite to form carbonic acid (H_2CO_3). When in contact with carbonic acid, minerals that contain lime, soda, potash, or other basic oxides are changed to carbonates. If carbonic acid is in contact with limestone ($CaCO_3$), for example, the weathered product will be calcium bicarbonate ($Ca(HCO_3)_2$) in a solution that may be lost through drainage.

Physical and chemical weathering do not occur in exclusion of one another. One group may be more active in a particular area, but both physical and chemical changes usually happen simultaneously. For example, a physical reduction in size hastens chemical activity by exposing a greater amount of surface area to attack. A one-inch cube of rock (or mineral) matter has a total surface area of six square inches. If the cube is broken into eight one-half-inch cubes, the area exposed to chemical weathering increases to twelve square inches (Figure 1.3).

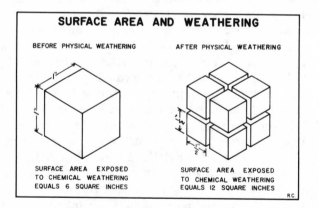

Figure 1.3 Relationship of the size of fragments to total surface area.

Another factor affecting weathering processes is that all minerals do not decay at the same rate. Some are more resistant than others and will have longer lives at the surface. Thus, under specified climatic conditions and types of parent material, the resultant soil should contain certain minerals in greater quantity than others.

Figure 1.4 illustrates the relative rate at which minerals in common igneous rocks chemically decompose. The resistance to decay reflects, to a large degree, the surface conditions under which the minerals experience weathering in relation to the conditions that existed when they originally solidified. Olivine, when forming from a molten state, crystallizes under high temperatures and high pressures. As a consequence, it tends to be extremely unstable when exposed to the low temperatures and pressures at the surface, and it weathers rapidly. Quartz, on the other hand, develops under considerably lower temperatures and pressures—in the late cooling stages of the magma—and it is relatively stable and very resistant to weathering.

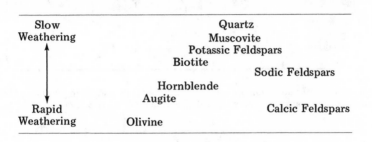

Figure 1.4   Resistance of minerals to weathering.

With a knowledge of the weathering sequence, it is possible to project the relative amount and type of mineral that will accumulate in the soil through time. Soluble materials such as calcium, magnesium, and sodium will generally be removed rapidly in humid regions where moisture is abundant. Under the same climatic conditions, oxides of silicon, iron, and aluminum—products of decomposition—are resistant and will most likely accumulate in the soil. A diagram illustrating the differing accumulations of mineral constituents is shown in Figure 1.5. Grouping the *bases* (those minerals containing calcium—Ca, magnesium—Mg, sodium—Na, and potassium—K) together, the diagram presents a hypothetical case of residual accumulation on a highly weathered tropical soil, and assumes a reduction in volume of the soil column of 50 percent.[2]

In a parent material containing equal quantities of the bases, the order of their weathering would be:

[2] A base is any molecule or ion that can combine with a proton. Normally within the field of soil science the cations of calcium, magnesium, potassium, and sodium are referred to as *exchangeable bases* or more simply *bases*. This terminology is employed because these four cations are associated with $CaCO_3$, $MgCO_3$, K_2CO_3, and Na_2CO_3 which are commonly occuring non-acidic compounds within the soil.

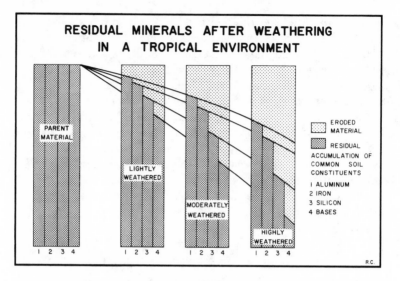

Figure 1.5 Accumulation of mineral constituents as weathering occurs. (The silicon is the mineral portion contained in silicate minerals; it is not quartz.)

Calcium	(Ca)	Greatest
Magnesium	(Mg)	↑
Sodium	(Na)	↓
Potassium	(K)	Least

While calcium is most susceptible to weathering, it is often found in the uppermost soil layer in slightly greater amounts than magnesium. The reasons for this are: (1) plants utilize more calcium than magnesium, therefore returning more calcium to the surface through plant absorption and release upon decomposing; and (2) calcium ions have a stronger affinity for binding to clay minerals than do magnesium ions.

The natural infertility of older tropical (and many other humid land) soils can be understood in the light of the weathering sequence in Figure 1.5. *Fertility* is "the status of a soil with respect to the amount and availability to plants of elements necessary for plant growth."[3] Man's food requirements mainly are provided by the grass family of plants, both for cereal grains and as forage for grazing animals. The grass family requires a relatively high availability of bases. Without bases—which as we know are highly soluble—grain and forage crops cannot thrive. Anyone familiar with farming

[3] *Glossary of Soil Science Terms* (Madison, Wisconsin: Soil Science Society of America, 1973), p. 7.

practices in much of the eastern United States is well aware that crushed limestone is frequently applied to fields. The limestone amends infertile soil by replacing calcium that was lost through drainage and plant removal, and it corrects the excessive acidity that limits the availability of other plant nutrients.

Products of Weathering

As time passes and the parent material undergoes continued weathering, a group of mineral fragments is produced from what was once the more massive original rock complex. Since a soil may be composed of particles of many sizes, the pedologist separates the fragments into groups for identification. The individual types are called *soil separates*. The name of each soil separate is shown in Table 2, along with its size range.

Sand grains, when dominant, yield a soil that can be easily worked by the farmer. Consequently, a sandy soil is said to be *light*. The major sand mineral is usually quartz (SiO_2), even though the coarser sands may contain rock fragments of varied composition (Figure 1.6). The dominance of quartz tends to make the sand fraction of the soil chemically inactive. Nutrients released by weathering are quickly leached away in humid climates, since sandy soils are typically highly permeable. Also, sands have very low water-holding capacity. So, although sandy soils are easy to work, they are limited for crop production by leaching, low-nutrient and water-holding capacities, and low natural fertility.

The chief function of sand is to serve as a framework for the more active soil components. Due to their relatively large size, sand particles provide the greatest degree of space between individual grains, thus promoting the movement of air and the drainage of water within the soil.

Silt, like sand, is composed dominantly of silicate minerals. It differs from sand mainly in its smaller particle size. Consequently silt has a total surface area per unit volume of soil greater than that of sand. It also has a faster weathering rate, releases soluble nutrients for plant growth more readily, and retains more "available" water than does sand.

Clay covers a wide range of substances with varied mineralogical and chemical characteristics. It is defined as mineral particles less than 0.00008″ (2 microns) in diameter, and generally falls into one of two broad groups—the silicate clays and the iron and aluminum hydrous-oxide clays. The silicate clay group is typical of weathering processes in the mid-latitudes, while the hydrous-oxide clays characterize types found in the tropics.

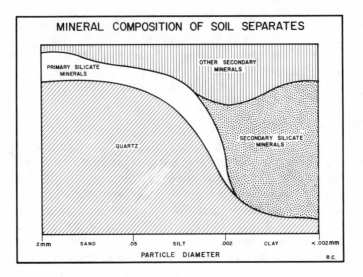

Figure 1.6 The relationship between particle size and the type of minerals present.

Table 2   Soil Separates

Soil Separate*	Diameter (mm)	Diameter (in)
Very coarse sand	2.00 - 1.00	.08 - .04
Coarse sand	1.00 - 0.50	.04 - .02
Medium sand	0.50 - 0.25	.02 - .01
Fine sand	0.25 - 0.10	.01 - .004
Very fine sand	0.10 - 0.05	.004 - .002
Silt	0.05 - 0.002	.002 - .00008
Clay	below 0.002	below .00008

*Stone, gravel, and organic material are listed separately and generally are not included in the fine earth analysis.

Most clay crystals resemble minute flakes in which atoms are arranged in a layered structure. The layers are of two forms: *silica sheets* and *alumina sheets*. A silica sheet is composed of tetrahedrons (pyramids), each having an oxygen atom at the vertices and a silicon atom in the interior. Each oxygen atom is equally spaced and equally distant from the silicon atom. By sharing oxygen atoms at the base of the pyramid, tetrahedra combine into hexagonal units. In repeating this pattern, these units form a *lattice* of the clay mineral.

An alumina sheet consists of octahedrons, each having a central atom of aluminum equidistant from six oxygen atoms or hydroxyls. The combinations of silica and alumina sheets relate to their stability and provide a basis for clay classification.

The very minute size of clay particles exposes an extremely large surface area (per unit volume) upon which chemical activity can take place. Unlike the more sterile sand separates, clay is an active portion of the soil—holding and exchanging ions.

Clay minerals have two principal sources. They can be formed directly by the alteration of parent material or indirectly by synthesis from weathering products, for example, crystallization. *Alteration* occurs when chemical weathering removes certain soluble components and substitutes others. *Recrystallization* is a complete change in the structure of the original minerals; for example kaolinite, the simplest of the clay minerals, may be formed from solutions containing soluble aluminum and silicon. Figure 1.7 shows the weathering sequence in the production of clay minerals. Various weathering stages are indicated. Chlorite and hydrous micas represent the youthful stages, and kaolin the old-age stage, of silicate weathering.

In contrast to sandy soils, which are considered light, clay soils are referred to as *heavy*. The very finely divided particles, packed closely together, create a dense soil of high potential plasticity and particle cohesion. When plasticity is high, dry soil may become hard and cloddy, and wet soil may be a sticky mass that can be molded or deformed by relatively moderate pressure. Plowing such a soil when wet can further reduce pore space, in some cases making the soil impervious to water and air. When the soil dries, it becomes very hard and dense. Soils with these characteristics are said to be *puddled.*

Although the soil separates—sand, silt, and clay—have been discussed here as individual entities, texture refers to the relative proportions of the separates in a given soil.

Texture is important for several reasons: (1) it influences the ease with which plant roots penetrate into the soil; (2) it affects aeration (exposure to air); (3) it influences moisture storage capacity; (4) it affects plant support capacity; and (5) it establishes the rate of chemical reactions within the soil.

For uniformity in soils description, the U.S. Department of Agriculture uses the percentages of the various soil separates in specific combinations. Each soil class is defined according to the relative proportions of the soil separates it contains (Figure 1.8).

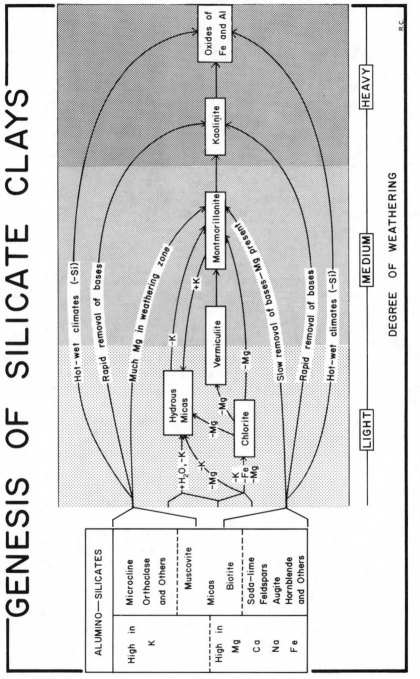

Figure 1.7 Conditions in which the various silicate clays and oxides of iron and aluminum may form.

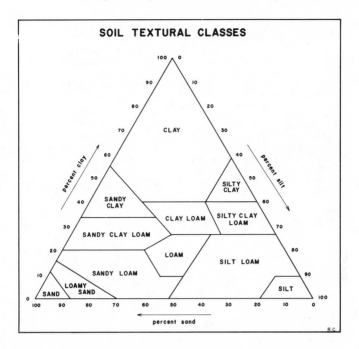

Figure 1.8 Chart showing the percentage of clay (below 0.002mm.), silt (0.002 to 0.05mm.), and sand (0.05 to 2.0mm.) in the basic soil textural classes.

Sands. Soil material that contains 85 percent or more of sand; the percentage of silt, plus 1½ times the percentage of clay, shall not exceed 15 percent.

Coarse sand. 25 percent or more very coarse and coarse sand, and less than 50 percent any other one grade of sand.

Sand. 25 percent or more very coarse, coarse, and medium sand, and less than 50 percent fine or very fine sand.

Fine sand. 50 percent or more fine sand or less than 25 percent very coarse, coarse, and medium sand, and less than 50 percent very fine sand.

Very fine sand. 50 percent or more very fine sand.

Loamy sands. Soil material that contains at the upper limit 85 to 90 percent sand, and the percentage of silt plus 1½ times the percentage of clay is not less than 15 percent. At the lower limit it contains not less than 70 to 85 percent sand, and the percentage of silt plus twice the percentage of clay does not exceed 30.

Loamy coarse sand. 25 percent or more very coarse and coarse sand, and less than 50 percent any other one grade of sand.

Loamy sand. 25 percent or more very coarse, coarse, and medium sand, and less than 50 percent fine or very fine sand.

Loamy fine sand. 50 percent or more fine sand, or less than 25 percent very coarse, coarse, and medium sand, and less than 50 percent very fine sand.

Loamy very fine sand. 50 percent of more very fine sand.

Sandy loams. Soil material that contains either 20 percent clay or less, and the percentage of silt plus twice the percentage of clay exceeds 30 percent, and 52 percent or more sand; or less than 7 percent clay, less than 50 percent silt, and between 43 and 52 percent sand.

Coarse sandy loam. 25 percent or more very coarse and coarse sand and less than 50 percent any other one grade of sand.

Sandy loam. 30 percent or more very coarse, coarse, and medium sand, but less than 25 percent very coarse sand, and less than 30 percent very fine or fine sand.

Fine sandy loam. 30 percent or more fine sand and less than 30 percent very fine sand, or between 15 and 30 percent very coarse, coarse, and medium sand.

Very fine sandy loam. 30 percent or more than 40 percent fine and very fine sand, at least half of which is very fine sand and less than 15 percent very coarse, coarse, and medium sand.

Loam. Soil material that contains 7 to 27 percent clay, 28 to 50 percent silt, and less than 52 percent sand.

Silt loam. Soil material that contains 50 percent or more silt and 12 to 27 percent clay, or 50 to 80 percent silt and less than 12 percent clay.

Silt. Soil material that contains 80 percent or more silt and less than 12 percent clay.

Sandy clay loam. Soil material that contains 20 to 35 percent clay, less than 28 percent silt, and 45 percent or more sand.

Clay loam. Soil material that contains 27 to 40 percent clay and 20 to 45 percent sand.

Silty clay loam. Soil material that contains 27 to 40 percent clay and less than 20 percent sand.

Sandy clay. Soil material that contains 35 percent or more clay and 45 percent or more sand.

Silty clay. Soil material that contains 40 percent or more clay and 40 percent or more silt.

Clay. Soil material that contains 40 percent or more clay, less than 45 percent sand, and less than 40 percent silt.

Soil Structure

Structure is the arrangement of soil particles into larger secondary units, called *peds*. Like texture, structure is an important physical characteristic of the soil. Although most soils are dominantly inorganic in composition, organic matter is extremely significant in the development of certain types of structures. Some soils, such as those made up entirely of sand, may show no structural arrangement due to the lack of binding elements. Other soils may have peds of varying shapes and sizes (Figure 1.9).

Soil structure is the result of many factors: the chemical nature of the clay, the amounts of clay and organic material in the soil, the nature of the soil's microorganisms, wetting and drying, freezing and thawing, and cultivation.

One theory to explain aggregation, or clustering, concerns the electrical charge of colloidal particles.[4] Water molecules are *dipolar*[5] in charge and can be firmly attached to colloidal nuclei; in fact, may serve as a link between two colloids. As water evaporates from the soil, the combining link is effectively shortened and draws together the colloidal particles and the larger soil grains to which they are attached. This process effectively clusters the colloids and larger soil

---

[4] Colloid refers to organic and inorganic matter having a very small particle size and a high surface area per unit mass.

[5] Dipolar means having two poles as a result of the separation of electric charge. A dipolar molecule orients in an electric field.

SOIL STRUCTURE: TYPES AND CLASSES

		PLATELIKE	PRISMLIKE	BLOCKLIKE	
Type of Ped					
		PLATY	PRISMATIC OR COLUMNAR	BLOCKY Angular or Subangular	SPHEROIDAL Granular or Crumb
Ped Size Classes	VERY FINE OR VERY THIN	I mm	I0 mm	5 mm	I mm
	FINE OR THIN	I – 2 mm	I0 – 20 mm	5 – I0 mm	I – 2 mm
	MEDIUM	2 – 5 mm	20 – 50 mm	I0 – 20 mm	2 – 5 mm
	COARSE OR THICK	5 – I0 mm	50 – I00 mm	20 – 50 mm	5 – I0 mm
	VERY COARSE OR VERY THICK	I0 mm	I00 mm	50 mm	I0 mm

J. Blok

Figure 1.9 The common manner in which soil separates are arranged into structural units and their size class.

particles. As evaporation continues and the colloidal material becomes further dehydrated, the particles may stick or cement into an aggregate.

Other significant factors include:

1. The *mycelial* (sort of cobwebby filaments) growth of microorganisms that serves as a mini-root system, binding particles together. Microbial gums may act as cements and can be the most important part of aggregate formation.
2. Water that has frozen in the pore space of soil exerts pressure on surrounding soil particles as the ice crystal expands. Pressing together stimulates the clustering of colloids and other soil particles. Furthermore, as moisture is withdrawn from the surrounding soil, the colloidal material is dehydrated—resulting in the cementing of particles in contact with one another. Thawing provides increase pore space in the areas previously occupied by ice crystals.

Structure is important because it changes the influence of soil texture on such factors are infiltration of moisture, aeration, the ability of plant roots to penetrate the soil, nutrient supplies, and the soil's resistance to erosion.

Micelle Activity and Plant Nutrient Availability

Earlier we stressed the importance of the clay fraction in soil separates. These particles, that are less than two microns (0.00008") in diameter, are chemically active and represent the plant's main reservoir for nutrients. Pedologists also recognize colloids within this category. As we know, colloids can be either organic or inorganic, but they must be less than 1 micron (0.00004") in size. For now we are concerned only with the inorganic colloids—also known as *micelles* (microcell).

Micelles, minute silicate-clay colloids, normally have a negative charge. Thus, they have the capacity to attract and hold positively charged ions, called *cations*[6] that have been released during the weathering of the parent material or added in fertilizers. Table 3 summarizes the products of the weathering of selected minerals and also lists the associated ions released during the weathering process.

Soilwater usually carries the cations to the micelle for adsorption.[7] As water percolates through the soil, it can transport dissociated ions and bring them in contact with the micelle (Figure 1.10). The adsorption of cations and their subsequent release is known as *cation exchange.* Each soil has a unique *cation exchange capacity* (CEC),[8] depending upon the existing amount of exchangeable ions. The quantity of exchangeable ions, in turn, is dependent upon: (1) the type of clay mineral, (2) the percentage of clay, and (3) the amount of organic colloids. In addition to a unique CEC, each soil carries a specific proportion of its cations as bases. The *percentage base saturation* (PBS) provides information on the relative amount of exchangeable bases—for example, calcium, magnesium, sodium, and potassium—to the total exchange capacity. In quantitative terms:

$$\text{Percentage Base Saturation (PBS)} = \frac{\text{Exchangeable Bases (milliequivalents)} \times 100}{\text{CEC (milliequivalents)}}$$

[6] As opposed to anions which have a negative charge.

[7] *Adsorption* refers to the adhesion of disassociated ions to soil colloids, whereas *absorption* involves ion assimilation.

[8] Cation-exchange capacity (CEC) is the sum total of exchangeable cations that a soil can absorb, expressed in milliequivalents per 100 grams of soil. An equivalent represents a quantity that is chemically equal to one gram of hydrogen. The number of hydrogen atoms in an equivalent equals 6.02×10^{23}. A milliequivalent is 6.02×10^{20} (0.001 gram of hydrogen).

Table 3   Weathering Products of Selected Common Minerals

Original Mineral	Products of Weathering Residual Mineral Products	Released Ions
Amphibole	Clay minerals, limonite, hematite	K, Ca, Mg, Na
Biotite	Clay minerals, limonite, hematite	K, Mg
Calcite	None for mineral; from limestones, some quartz, clay minerals, and hematite as impurities.	
Chlorite	Clay minerals	Mg, Fe^{++}
Clay minerals	Under high moisture in tropics may develop bauxite	SiO_2
Dolomite	None for mineral, from dolomite rock some quartz, clay minerals, and hematite as impurities.	Ca, Mg, HCO_3
Garnet	Garnet	
Gypsum	None for mineral; from gypsiferous shale or sandstone, clay minerals and quartz.	Ca, SO_4
Hematite	Hematite	
Limonite	Limonite	
Muscovite	Muscovite tends to remain, eventually alters to clay minerals and quartz.	K, SiO_2
Olivine	Clay minerals, limonite, hematite	Mg, Fe^{++}
Orthoclase & Microcline	Clay minerals, quartz	K, SiO_2
Plagioclase	Clay minerals	Na, Ca
Pyroxene	Clay minerals, limonite, hematite	Mg, Fe^{++}, Ca
Pyrite & Marcasite	Limonite, hematite	Fe^{++}, SO_4
Quartz	Quartz	Some SiO_2
Serpentine	Serpentine	
Talc	Talc	

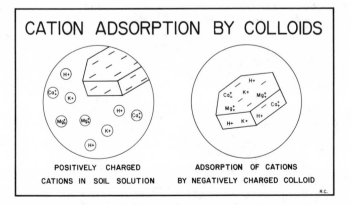

Figure 1.10   Diagram of a clay colloid (micelle) with its sheetlike morphology, numerous negative charges, and swarm of absorbed cations.

The base saturation of soils in arid climates—where moisture is low and the soluble bases are not leached out[9]—tends to be high. In humid climates, PBS is generally lower because the greater amounts of water percolating through the soil remove the bases. The importance of percentage base saturation is directly related to the availability of the bases in meeting the nutrient needs of plants; that is, as an index of fertility.

One particular cation, hydrogen(H^+), in appreciable quantity on the micelles produces soil acidity. An absence of H^+ and a high amount of basic cations produce an alkaline situation. The degree to which the H^+ ion is present in the soil is a measure of nutrient availability, because when hydrogen cations dominate the micelles they displace the bases and subject them to removal. The dominance of H^+ in some soils is due to the chemical force with which ions are held. In general, the following order represents the strength of ion attachment to colloids and the order of ion accumulation in humid area soils:

$$H > Ca > Mg > K > Na.$$

Pedologists have devised a measure of H^+ concentration known as the pH. Most soils vary in pH from about 4 to 10, and are distinguished from one another by their positions on the pH scale (Figure 1.11). Each unit change on the scale represents a tenfold change in the concentration of H^+. A pH value of 7 indicates neutrality—for example, H^+ and OH^- ions in equal concentrations.[10]

[9] *Leaching* is the removal of soluble materials in solution from the soil.
[10] The neutral point (pH of 7) is established for the concentration of H^+ ions in pure water at a temperature of $75°F$. This amounts to 1.0×10^{-7} grams of H^+ ions per liter of water and is also equal to the number of OH^- ions present. Since very small numbers, such as 1.0×10^{-7}, or 0.000,000,1, are difficult to visualize, the pH expression is given as the negative logarithm of the hydrogen activity of a soil.

pH	ACIDITY	
	normality of H^+	exponential expression
0	1.0	1.0×10^{0}
1	0.1	1.0×10^{-1}
2	0.01	1.0×10^{-2}
3	0.001	1.0×10^{-3}
4	0.000,1	1.0×10^{-4}
5	0.000,01	1.0×10^{-5}
6	0.000,001	1.0×10^{-6}
7	0.000,000,1	1.0×10^{-7}
8	0.000,000,01	1.0×10^{-8}
9	0.000,000,001	1.0×10^{-9}
10	0.000,000,000,1	1.0×10^{-10}
11	0.000,000,000,01	1.0×10^{-11}
12	0.000,000,000,001	1.0×10^{-12}
13	0.000,000,000,000,1	1.0×10^{-13}
14	0.000,000,000,000,01	1.0×10^{-14}

Knowing the pH of a soil is extremely important to insure proper soil management. For example, the lower the pH, the greater the acidity and decreased availability of base nutrients. In such a case an application may be necessary to supply the soil with calcium and to raise the pH to a desirable level.

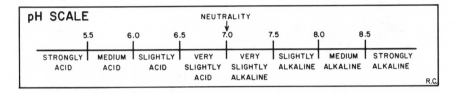

Figure 1.11   The pH scale to measure the concentration of the hydrogen cations (H^+) in the soil.

Soil Color

Color is a necessary part of any discussion concerning the parent material's influence on soil characteristics. The color of a soil, one of its most readily observable features, is the result of various chemical weathering processes acting upon parent material, coupled with organic material. (Organic material, which will not be discussed here, can be the dominant coloring agent in the soil—the carbon content gives it a dark color.)

The most important inorganic coloring agent is iron (fe). Iron compounds in the parent material produce a variety of colors:

1. Under poor drainage conditions, where oxygen is deficient, reduced iron yields gray to bluish-gray colors.
2. Sites of good aeration and drainage lead to the oxidation of iron, producing red colors.
3. Where the soil is moist a great deal of the time, the iron may be hydrated as well as oxidized, resulting in a yellow color.

Other soils, especially in arid regions, may have accumulations of mineral salts. These soils often show a surface mineral encrustation that is whitish in color.

Obviously, color provides a clue to drainage qualities and the sort of parent material from which the soil developed. However, color is only a *clue*; it is not the complete answer to soil characteristics and processes. For example, a red shale or sandstone, when weathered, may yield a red-colored soil even though the oxidation of iron is not the primary process.

The Organic Fraction of the Soil

2

The organic portion of the soil is extremely varied and complex, acting as both a substance and as an agent in decomposition. Organic matter averages only about five percent by weight in an ideal loam, but its role in influencing soil properties and as a fundamental part of the biochemical activities taking place within the soil is of great importance.

SOIL ORGANISMS

From microscopic single-celled organisms to large burrowing animals, the soil seethes with life. It is anything but an inert substance, as an examination of the often unseen plant and animal inhabitants would prove. Within a single cubic centimeter of fertile topsoil, over 1,000,000,000 bacteria may be present; under normal conditions in a single acre protozoa may number as many as 1,000,000, and earthworms may range from 250,000 to 1,000,000. Yet, these life forms represent only a few of the soil's many "tenants."

Microorganisms by far outnumber the larger plants and animals of the soil. The reason is that the soil provides them with an amenable habitat, plus food and water. Unlike higher plants, which can change energy into food through photosynthesis, microorganisms primarily obtain their energy from the tissue of the higher plants. In the process of decomposing plant tissue (for example, organic residue), the microbes are capable not only of obtaining energy, but also of simultaneously releasing back to the soil the many nutrients existing as complex organic compounds within the plant. This is an extremely important natural process that permits a recycling of plant nutrients. If, for example, a given stand of plants absorbs most of the available nutrients in the soil, future plant growth would be extremely limited due to the diminished food supply. However, microorganisms, through their feeding habits, make previously absorbed plant nutrients once again available as soluble inorganic compounds. Some of the more important soil organisms are shown in Figure 2.1 and are briefly described in the following pages.

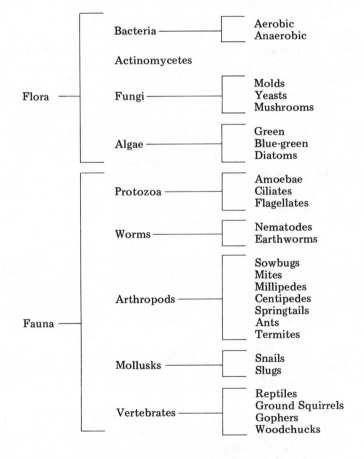

Figure 2.1   Some important soil organisms.

Soil Flora

Plant life in the soil varies both in form and function. Sizes range from the large roots of trees to forms that are less than 1 micron (0.00004″) in length, and their number is highly dependent upon living conditions. The most numerous, by far, are the smallest and most simply structured floral groups (*microflora*), which can either be increased or limited in number by the nutrient supply, temperature, moisture, or pH level of the soil environment.

Certain types of microbes obtain energy by oxidizing simple inorganic substances. Most, however, derive their primary sources of energy and their essential elements from the organic materials in the soil. While they may carry on decomposition over a wide temperature range, warmer regions usually encourage greater microbial activity. The maximum efficiency in decomposition is estimated at approximately 95°F. Moisture and aeration, closely associated, are also critical to the type and number of flora that are present. For *aerobic* (free oxygen dependent) populations, the most desirable soil moisture is approximately 50 to 70 percent of "field capacity." Water logging and deficient atmospheric gases can create a habitat suitable only to *anaerobic* (not dependent upon free oxygen) life forms. Also, certain soil organisms cannot tolerate acid or alkaline extremes; for example, acid soils are favorable for the growth of fungi but unfavorable for the development of legume bacteria, nitrifying organisms, and actinomycetes. Each type of soil occupant has a unique set of habitat requirements that enables it to perform specific and necessary tasks, normally those of decomposing organic residues and recycling mineral nutrients.

Bacteria. Bacteria are the simplest structured and the most numerous life forms existing in the soil. Functioning as decomposing agents and—in many instances—nitrogen fixers, they might be classified as one of the most important groups of soil inhabitants. As single-celled organisms they are able to achieve dense populations in rather short periods by elongating and dividing into two parts. This microfloral group can be subdivided into aerobic and anaerobic forms. Another subdivision can be made according to their energy source. *Autotrophic* bacteria obtain energy by oxidizing inorganic material; *heterotrophic* bacteria receive energy by decomposing organic materials.

The average-sized bacteria is 1 micron (0.00004″) long by ½ micron (0.00002″) in diameter and usually functions best under oxygen and moisture conditions that are considered most desirable for higher plants. While capable of surviving under a wide range of temperatures, their activity is greatest between 70°F and 100°F.

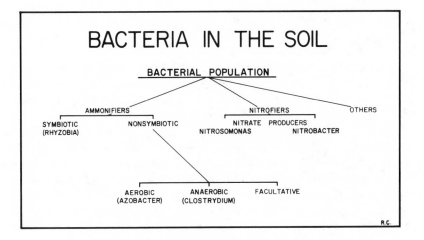

Figure 2.2   Selected bacterial groups and their functions.

Certain bacteria groups can function at low pH status, while others perform their work efficiently at high levels. Most, however, function best at pH values near neutrality—about 6 to 8—and when there is a sufficient amount of exchangeable calcium in the soil.

In addition to decomposing organic residue, probably the most significant role of soil bacteria concerns their efficiency in immobilizing nitrogen to a form in which higher plant life can assimilate it. Nitrogen is essential for a good rate of plant growth and respiration. When it is not present in sufficient quantities, the plants' ability to utilize basic nutrients is severely limited, plant growth is stunted, and leaves lose their deep green color, becoming pale and yellow due to a loss of chlorophyll.

Nitrogen fixation (immobilization), although not confined solely to bacteria, involves the efforts of numerous members of this microfloral group—the ammonifiers (ammonia-producers) and the nitrifiers (nitrite and nitrate-producers). Figure 2.2 illustrates the interrelationships and functions of some of the more important soil bacteria.

Most of the bacteria in the soil require nitrogen either in mineral form, such as ammonium and nitrate, or as organic nitrogen compounds, such as plant and animal proteins. Very few microflorals are capable of utilizing nitrogen gas in its atmospheric free state. One of the few soil bacteria that can do so is the genus *Rhizobium*. *Rhizobium* are known as *symbiotic* bacteria; that is, they live in a mutually beneficial relationship with other organisms—normally with the leguminous family of plants. These bacteria infect the root system of plants, develop nodules as they grow, and in a parasitic fashion derive food and minerals from their hosts. In turn, *Rhizobium* are able to utilize atmospheric nitrogen and also to transfer it to the plant.

Among the *nonsymbiotic* (independently existing) bacteria, the most significant in terms of assimilating free nitrogen into their cellular structures are the genus *Azobacter* and the genus *Clostridia*, both heterotrophic. The specific members dominating the *rhizosphere* (root zone) are dependent upon habitat status. Under conditions of good aeration, high organic content, neutral or alkaline reaction (pH $\geqslant$ 6.0), and available calcium, the aerobic *Azobacter* are most numerous. (Soil pH seems to be their most restrictive growth factor.) *Clostridia*, on the other hand, are anaerobic—perhaps even *facultative*, meaning they can exist in either an aerobic or anaerobic state. They can withstand highly acidic conditions and are spread over a wide geographic area. Since even an aerated soil may have either water pockets or saturated pore spaces, *Azobacter* and *Clostridia* can operate within the same soil simultaneously.

Bacteria's role in making nitrogen available to higher plants can be better appreciated when it is realized that most of the nitrogen stored within the soil is associated with organic matter. Soil organic matter generally contains 5 to 6 percent nitrogen, most of which is immobilized in protein form—primarily as amino compounds—and is unavailable to higher plants. In the process of digesting organic matter to obtain energy for growth, heterotrophic bacteria set free ammonia (NH_4) as a byproduct. This process is known as *ammonification*. The released NH_4 may then: (1) be reappropriated by bacteria, (2) be utilized by other microorganisms, for example, mycorrhizal fungi, (3) be made available to higher plants, (4) become fixed by clay minerals and organic matter, or (5) become nitrified.

The *nitrification* process is the work of another group of special purpose autotrophic bacteria known as *nitrobacteria*. Within this group are *nitrosomonas*, which convert ammonia into *nitrites*, which are in turn oxidized to *nitrates* by *nitrobacter*. The end product—nitrate nitrogen—is in a stable form and can be readily assimilated by plants.

Nitrogen fixation and recycling are possibly the most important roles of soil bacteria, but other bacterial functions are also of major significance. In the process of digesting plant residues, the microbial population temporarily immobilizes carbon, energy, and numerous nutrients. These nutrients once again become available for mineralization and plant use upon the death and decomposition of the microbials. The net effect is not only the recycling of nitrogen, but also of sulpher, phosphorus, calcium, magnesium, potassium, and other nutrients. Hence, these minute and unseen plants, at times numbering in the millions per gram of soil, function as efficient converters of once used materials into forms that again are available for higher plant life.

Actinomycetes. Only bacteria are more numerous in the soil than these organisms, which display morphological features of both bacteria and fungi. Actinomycetes are one-celled organisms of approximately the same characteristics in cross section as bacteria, but they also possess long, threadlike, branched filaments. Because of their appearance they are sometimes referred to as "ray fungi." Their total population is approximately 10 to 25 percent less than bacteria, but due to their larger size and filaments their gross organic weight in a unit volume of soil is about the same as that of bacteria.

The primary activity of actinomycetes centers on the decomposition and *humification*[1] of organic residues. They are extremely active (and most numerous) in organic material that is in the later stages of decay, and are especially important in the further breakdown of the more resistant organic compounds. These microbes can function in soils of low moisture and have the greatest moisture-range tolerance of any of the microflorals, as long as available calcium is not restrictive. Actinomycetes prefer soils ranging in reaction from 6.0 to 7.5 and are very sensitive to acidity— normally they do not exist in soils at a pH of 5.0 and lower.

In addition to their role in decomposition, certain species of actinomycetes produce antibiotic substances of value as medicines, while others cause potato scab and pox in sweet potatoes. The devastating potato famine in Ireland during the 1800s was probably due to soil infestation by actinomycetes.

Fungi. The fungi floral class is generally subdivided into three major groups: mushroom fungi, molds, and yeasts. (The last group is very limited and not of importance in most of the earth's soils.) Fungi are heterotrophic plants, ranging in size from microscopic molds to the large and complex fruiting mushrooms and bracket fungi. Of the bacteria, actinomycete, and fungi population, fungi account for probably no more than 1 percent of the total,[2] even though in actual amount of cell substance the combined weight of the fungi group may equal or exceed that of bacteria. This is a result of the group's high degree of efficiency in transforming the

[1] Humification is a process or condition of decay in which plant or animal remains are so thoroughly decomposed that their initial structures or shapes can no longer be recognized.

[2] Francis E. Clark, "Living Organisms in the Soil," in *The 1957 Yearbook of Agriculture: Soil* (Washington, D.C.: U.S. Govt. Ptg. Office, 1957), p. 159.

substances which they attack into their tissues. It is estimated that as much as 50 percent of the organic wastes decomposed by molds may become fungal tissue.[3]

Fungi perform several important functions including: (1) decomposition of resistant organic materials, (2) extending the effective rooting system of some plants, and (3) the production of antibiotic substances. Molds and mushrooms are capable of vigorous activity in soils from acid through alkaline reactions. Thus, in highly acid soils where the growth of bacteria and actinomycetes is restricted, fungi may become the dominant decomposer of organic residues. The persistance and vigor of these plants enable them to decompose even the very resistant lignin.

Some fungi can extend the effective feeding area of specific plants in a "fungus root" association known as *mycorrhizae*. These associations are most common in forested areas and appear to be of mutual benefit to the plant and the fungi. Since the *mycelial*[4] threads of the fungi have a tremendous surface absorbing area, the fungi in contact with the root system act as an extension, increasing the soil contact area several hundredfold and adding to the food supplies of the plant. One gram of soil may contain 10 to 100 meters of mold filament. The fungi, on the other hand, can meet their food requirements by absorbing carbohydrates from plant roots. This process is extremely important for the growth of certain types of forest species in nutrient deficient soils, where a mature forest association could not develop without mycorrhizae.

Certain groups of fungi have received much attention due to their production of antibiotic substances. The best known is *penicillium*. Soil research in the Soviet Union has revealed that species of penicillium amount to approximately half of the total fungi population in most Russian soils, about 30 percent in semiarid, and 21 percent in desert soils.[5]

As we know, fungi are capable of vigorous growth under a wide range of environmental conditions. The most restrictive factor affecting their growth is the availability of oxygen.

Algae. Algae are simple structured, chlorophyll-bearing organisms that are capable of carrying on photosynthesis in a manner comparable to the higher plants. Although they are the highest form of soil microflora, their total number is only a small fraction of the

---

[3] Harry O. Buckman and Nyle C. Brady, *The Nature and Properties of Soils* (New York: Macmillan Co., 1971), p. 122.

[4] Mycelia are the threadlike bodies of simple organisms, such as the common bread mold.

[5] B.T. Bunting, *The Geography of Soil* (Chicago: Aldine Publishing Co., 1967), p.41.

microbial population. Some algae forms exist well beneath the soil surface by using plant residues or soil organic matter for their food supply. Most species, however, must live close to the surface in the presence of light, which supplies the energy needed to combine carbon dioxide and water for the manufacture of carbohydrates.

The common forms in which algae occur in the soil are: (1) blue-green, (2) green, and (3) diatoms, the least numerous. Blue-green algae can fix nitrogen to form protein, especially in water-logged soils (such as rice paddies). Having a worldwide distribution, algae are among the first occupants of exposed rock material, aiding in the formation of an initial organic layer in the process of soil development, but their role in contributing organic matter and/or altering the characteristics of mature soils seems to be of minor importance.

Soil Fauna

The soil's animal population is no less varied in form and function than are the flora members. Here, too, the most numerous are the most minute. Whether these animals are micro or macro in size, however, they are all involved in the task of altering certain aspects of the soil environment. The larger inhabitants are often quite effective in changing the soil's physical characteristics. Through digging or burrowing they may pulverize, mix, or transport earth materials both vertically and horizontally, and in the process may help to increase aeration and drainage and/or alter soil structure. All soil fauna are capable of translocating organic matter and altering it chemically through digestion; thus hastening decomposition.

In combination, soil animals represent an organic component of the soil. When their life cycles are ended they become a substrate for decompositional activity. One earthworm might not seem to be a significant organic contribution, but as many as 1,000,000 earthworms may be found in one acre of soil, and their total body weight may be more than 1,100 pounds. Individual size cannot be the sole factor in assessing the impact of specific fauna in the soil.

Protozoa. The single-celled protozoa are the most profuse members of the animal kingdom in the soil, numbering over 1,000,000 per soil gram. Their structure is more complex and their size slightly larger than bacteria.

Protozoa are divided into three broad groups based on morphological features: (1) amoeba, (2) ciliates, and (3) flagellates. As the names suggest, ciliates have *cilia* (hairs) and flagellates possess whiplike appendages known as *flagella.* Flagellates are the most numerous of the protozoa, followed by amoeba and then ciliates.

The protozoa live in aquatic environment—in the films of water surrounding individual soil particles. If food supplies become limited or if the soil dries out, these animals are at le to *encyst* (enclose as if in a capsule) and later reactivate when conditions are more favorable. Their feeding habits are varied. Some feed on bacteria and in the process stimulate the nitrogen-fixers of that group. Others feast on fungi, algae, or dead organic matter. Protozoa effectively hasten decomposition by chemically altering organic components of the soil through enzymatic digestion. Although these organisms do have a predatory nature, they do not seem to interfere with higher plants.

Worms. Nearly everyone is familiar with the common earthworm. Yet, far more numerous in the soil are the microscopic and nonsegmented nematodes, also known as threadworms or eelworms. Billions of them may be found in the surface layer of one acre of cropland. Nematodes are classified according to feeding habits. One group feeds on decaying organic material; a second is predatory, feeding on other nematodes, earthworms, bacteria, and protozoa; and a third is parasitic, infesting the roots of higher plants. Nematodes can be both beneficial and harmful. As decomposing agents, they bring about a mixture of mineral and organic matter. When too abundant, however, the parasitic group can cause considerable damage to plants by infesting roots and by increasing the susceptibility of the plant to attack by other parasites.[6]

Due to widespread distribution, careful land-management practices must be exercised to avoid a concentrated population of nematodes. An example of how difficult this can be is shown in the problem plaguing tobacco farmers in eastern North Carolina. Tobacco is a crop especially susceptible to attack by the "root knot," "meadow," and "stunt" members of the parasitic nematodes. In an effort to reduce their numbers, farmers often rotate tobacco with crops that are unfavorable hosts for these parasites. The complexity of the problem is compounded, however, by the fact that among the nematodes are species and strains that differ in crop preferences. A corn-tobacco rotation, for example, gives the tobacco protection from certain "root knot" strains that cannot thrive on corn, but corn is a favorable host for the "meadow" and "stunt" members, which will increase in numbers and infest the tobacco when planted. Hence, although a specific crop rotation may reduce the number of some parasitic species and strains, it may enhance the growth potential for others. A rotation of oats followed by weeds often gives fairly good control of most kinds of nematodes.[7] Proper

[6] In the process of penetrating the plant root, nematodes open entrances into the plant for other less efficient parasites.

[7] W.E. Colwell, "Tobacco," in *The 1957 Yearbook of Agriculture: Soil*, p. 656.

rotation, coupled with fumigation, may be necessary in many instances to control the adverse effects of these soil microbes. This involves a significant investment of both time and capital.

Considerably larger in size than the nematode, and probably the most well-known soil inhabitant, is the common earthworm. Thorp notes that the sizes of earthworms vary to a considerable extent, some species being as long as 18″ and up to ½″ in diameter.[8] Ingesting organic matter for food and inorganic matter by burrowing, they chemically alter the soil as these materials pass through their digestive systems. The amount thus changed may total 20 tons of soil per acre per year.[9]

The most important effects of earthworm activity are: (1) the vertical and horizontal mixing of soil, and (2) the increased aeration and drainage resulting from burrowing. A widely held opinion is that earthworms increase soil fertility, but there appears to be little evidence to support this claim. These macroorganisms tend to be restricted by low soil reaction, hence are more abundant in soils with higher amounts of available calcium. Therefore, they can be considered as an indication, rather than as a cause, of soil fertility. In the process of digesting organic matter, humification takes place in the worm's stomach before the material is excreted. Altered in size and transported to lower soil levels, the material most probably is reduced to colloids and serves as an ion absorber in the cation-exchange processes in the soil.

Arthropods. The number of arthropods in one acre of well manured land has been estimated at 7,720,000. Although the individual insects are very small and provide little organic material, their short lives, rapid reproduction rate, and vast numbers make them a significant feature of the soil. Not only do they constitute a part of the organic soil fraction, but they also can materially affect the mineral content of varied levels of soil through mixing. Thorp gives a succinct description of soil alteration by the leaf-cutting ants of the tropics:

The leaf-cutting ants march for long distances, cut out fragments of leaves and stems of plants that are to be used directly or indirectly for food, carry them back home, and store them in underground chambers. Here they are used to produce fungus that is in turn used for food. In this way organic matter, both in the forests of Central America and in the brushlands of

[8] James Thorp, "Effects of Certain Animals That Live in Soils," *Selected Papers In Soil Formation and Classification* (Madison, Wisconsin: Soils Science Society of America, 1967), p. 195.

[9] Ibid., p. 164.

northern Mexico and Texas, is incorporated in the soil and converted to humus through the activities of fungi (planted by the ants) and by the deposition of fecal matter by the ants themselves. In a private communication, R. L. Pendleton reports that mounds of these leaf-cutter ants in Central America frequently are as much as 15 feet across and 3 feet high. I have seen smaller ones in the brushy lands of south Texas. The mere tunneling operations of these ants have the effect of moving mineral material from one horizon of the soil to another, and after the anthills are abandoned the chambers provide channels for rapid water penetration to the deeper subsoil horizon.[10]

Another example of the arthropod's impact on the soil is demonstrated by the role of termites in the tropics. Termite mounds can average from 6 to 10 feet in height and may be up to 50 feet in diameter, with as many as 30 mounds per acre. Such construction amounts to transporting an estimated 10,000,000 pounds of soil per acre.[11]

Other arthropods include the *mites* (microscopic to barely visible in size) and the *springtails*. Mites may reach a population of several billion per acre. Some types feed on other small soil fauna, but most feast on organic matter—as do the springtails—and contribute significantly to the breakdown of organic residues and fungi.

Other soil fauna.   Numerous larger animals also find the soil a suitable habitat. Mollusks (snails and slugs), crustaceans (crabs and crawfish), reptiles (snakes and lizards), and mammals (gophers, ground squirrels, woodchucks, and other small vertebrates) are all effective in the total process of granulating, pulverizing, and transferring large amounts of soil, and incorporating organic matter. Furthermore, by burrowing they normally contribute to an increase in soil aeration and drainage.

A Summary of Soil Inhabitants

The life cycle of soil organisms may be said to interlock with that of higher plants, each group requiring the other for survival. Habitat factors are extremely important to the microflora and fauna. To develop and grow, they must have an energy input, as well as many of the elements essential to higher plants. Energy is obtained either through the oxidation of inorganic substances (autotrophic

[10] Ibid., p. 196
[11] Clark, *Living Organisms,*" p. 164.

bacteria) or from organic materials (heterotrophic bacteria, fungi, actinomycetes, and the like). The organic remains of higher plants are the prime suppliers of the basic needs of most soil life forms. The decomposition of organic matter releases nutrients previously held in tissues, making them available for further plant growth. This prevents the continued accumulation of vegetative debris. During the process, certain microbes—for example, bacteria—are able to fix free nitrogen. This capability cannot be underestimated, because in its elemental form, nitrogen—necessary for good plant productivity—is unavailable.

The soil's organisms also have other requirements beside energy and nutrient supplies. Some of the important ones that may affect the size of soil populations are: aeration, moisture availability, pH (the degree of acidity or alkalinity), and temperature.

Microorganisms also may significantly influence the stability of soil aggregates. During intermediate stages of decomposition, microbial gums and slimy vegetative products can cause soil particles to stick together and increase their resistance to breaking apart. Not all activities of the soil inhabitants, however, are beneficial. Crops may be destroyed by plant root infestations or burrowing animals, and soil organisms may compete detrimentally with higher plants for available nutrients and oxygen, if they are limited.

SOIL ORGANIC MATTER

Organic matter refers to all living and dead matter within and upon the soil. The organic matter consisting of nondecomposed leaves, twigs, or stalks on the surface is known as *litter;* the plant and animal residues in the process of decomposition are called *humus.*

The amount of organic matter in the soil is determined by a combination of several factors:

1. *Climate and Vegetation.* The distribution of world vegetation types is closely associated with specific patterns of temperature and precipitation. It is common knowledge, for example, that the sparse vegetation of desert regions is intimately tied to the meager precipitation. In transitional climates, between desert and forested environments, a precipitation gradient exists—increasing in the direction of the more humid forested areas. The native vegetation of this region is dominantly of the grass family. The height and completeness of the surface cover, and the luxuriousness of the grasses, increase in the direction of increasing precipita-

tion (Figure 2.3). Similarly, as grass cover and its associated root complex become more abundant, the organic content of the soil increases. In the grasslands of the United States this variation ranges from 80 tons of organic matter per acre—within the top 40″ of the soil—along the arid climatic boundary, to 160 tons per acre within the humid, tallgrass prairie region.[12] When precipitation exceeds the environment's moisture requirements,[13] a forest cover of vegetation is usually present. The organic matter content of forest soils is less than in the grasslands. The reason lies in the structure of vegetation. The root system of grasses is concentrated near the surface and diminishes gradually with depth, providing the greatest organic content near the surface and gradually decreasing downward. The roots of trees, on the other hand, are deeply distributed, but trees supply most of their organic debris as leaf fall, concentrating the major portion of plant material on the surface rather than incorporating it within the soil.

Another climatic influence on the accumulation of organic matter relates directly to temperature. In cold climates where soils are frozen for several months of the year, the rate of decomposition is extremely slow and organic matter accumulates. Consistently warm and moist climates favor the development of active microbial populations that decompose material throughout the year, tending toward low soil organic content (Figure 2.4).

2. *Parent Material.* Inorganic materials can influence the amount and type of vegetation growing on a particular site. Minerals that supply abundant nutrients, in particular calcium and phosphorus, foster more vigorous vegetative growth than that which occurs on sites deficient in plant foods. Because the profuseness of vegetation varies, the potential for the accumulation of organic matter in the soil varies also.

3. *Topography.* The affect of surface relief on organic matter accumulation is primarily related to drainage conditions.

---

[12]H.D. Foth and L.M. Turk *Fundamentals of Soil Science*, 5th ed. (New York: John Wiley and Sons, Inc., 1972), p. 132.

[13]An area's moisture requirement is generally considered a function of the energy available for evaporation and transpiration processes. This "moisture demand" is called *potential evapotranspiration*, and is defined as the total amount of moisture which could be evaporated and/or transpired under conditions of optimal soil moisture.

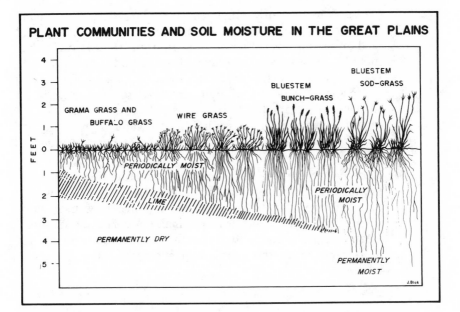

PLANT COMMUNITIES AND SOIL MOISTURE IN THE GREAT PLAINS

Figure 2.3 The relationship of plant communities to the depth of soil moisture penetration in the Great Plains of North America. *(After Shantz: Courtesy of the Association of American Geographers.)*

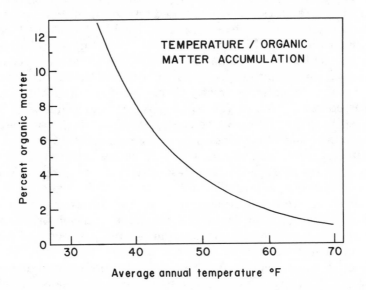

Figure 2.4 Organic matter accumulation bears an inverse relationship to temperature. Organic matter decomposition rates, however, increase with increased temperature.

Depressions toward which moisture drains will normally have greater organic matter accumulations than adjacent well-drained slopes. Factors contributing to this difference include: (1) greater availability of moisture for plant growth, (2) downslope surface wash of fine-sized organic matter, (3) downslope migration of colloidal-sized organic material in suspension within the soil, and (4) decreased oxidation of organic matter in depressions due to water dominating the pore spaces for greater lengths of time than on adjacent slopes.

4. *Time.* A vegetation cover and the accumulation of organic matter are not instantaneous occurrences. Beginning with a bare surface and lacking nitrogen, plant growth is slow. Therefore, organic matter is minimal. Only through time and weathering do nutrients become available to provide for more rapid vegetative growth and increased organic accumulations.

Humus

Humus is a dark colored (usually brown or dark brown), rather stable fraction of soil organic matter that remains after the major portion of residues has decomposed. As we mentioned earlier, the processes involved in the decomposition and leading to the formation of humus are called humification. An interesting aspect of humus development is that organic residues are broken down selectively by soil organisms. Thus, as decomposition takes place, the composition of the organic matter is greatly changed from its original form. Plant tissue contains cellulose (20 to 50 percent), hemicellulose (10 to 30 percent), lignin (19 to 30 percent), protein (1 to 15 percent), and fats, waxes, and other substances (1 to 8 percent). In the formation of humus, the celluloses and hemicelluloses are reduced rapidly. Lignin, being more resistant to destruction, is chemically altered, but remains in the soil as the dominant residual. Proteins are also retained in significant quantities, but for different reasons. It is thought that protein molecules are either: (1) absorbed on the surface of clay minerals and become resistant to decomposition, or (2) protein decomposing enzymes are trapped by clay particles, rendering them inactive. The result of such preferential decomposition is a soil organic content with percentage remains as follows: cellulose (2 to 10 percent), hemicellulose (0 to 2 percent), lignin (35 to 50 percent), protein (28 to 35 percent), and fats, waxes, and others (1 to 8 percent).

Humus produces significant changes in the soil complex. As humus increases:

1. The soil's ability to hold exchangeable ions increases. Humus of colloidal size (less than 1 micron) has a high cation-exchange capacity—approximately twice that of clay.[14] Cations such as Ca, Mg, and K are thus prevented from leaching and are maintained in a form available for microorganisms and higher plants.

2. The soil's water-holding capacity is increased. Humus has the ability to absorb large quantities of water, shrinking and swelling with moisture changes.

3. The soil is provided with an important source of carbon and nitrogen. The carbon content of humus is theoretically averaged at 56.24 percent, and the nitrogen content at 5.6 percent. The ratio of these two is 10.04:1, which approximates very closely the ratio found in many soils. Determination of the *carbon-nitrogen* (C-N) *ratio*, which is rather constant for soils within a given climatic area, provides an index of the breakdown rate of organic matter by microorganisms. Microbial enzymatic processes convert organic carbon to gaseous carbon dioxide to obtain energy, yet these organisms are inefficient in incorporating the carbon into their cell structure. As decomposition proceeds, considerable carbon is lost to the atmosphere, whereas practically all of the nitrogen is incorporated into new protein molecules, resulting in a narrowing of the carbon-nitrogen ratio.

   The practical significance of the C-N ratio comes from the strong competition for nitrogen between higher plants and the microbials. When organic residues with a wide C-N ratio are incorporated within the soil, microbial activity is stimulated, soon reaches its peak, monopolizes the nitrogen, and will produce a nitrogen deficiency for plants.

4. The *tilth* of soils is improved. Tilth refers to the physical conditions of the soil as related to its ease of tillage, fitness as a seedbed, and its resistance to emerging seeds and root penetration.

5. Soil structure is improved. Active microbial activity associated with humus production results in mycelial growth and microbial gums capable of binding soil separates together.

[14]Louis M. Thompson, *Soils and Soil Fertility* (New York: McGraw-Hill Book Co., 1951), p. 49.

ORGANIC MATERIAL AND SOIL COLOR

In Chapter I, soil color was considered solely in terms of the chemical weathering of inorganic materials. Organic matter, however, can contribute to the color of a soil and can be sufficiently dominant to mask the inorganic coloration. A high content of organic matter existing in pore spaces and as coatings on mineral particles may give the soil a dark or black color. Prairie soils, where luxurient growths of grasses occur, are very dark colored. Low organic content, on the other hand, leads to lighter coloration, typical of desert soils.

Soil Porosity, Moisture, and Atmosphere

3

An ideal loam soil, as discussed in Chapter 1, contains approximately 45 percent mineral matter. The remaining portion consists of about 5 percent organic matter and 50 percent pore space—about 25 percent each of air and water. Air and water are extremely variable elements in the soil complex, the quantity of one varying inversely with the other. When the soil is saturated with water after a heavy rain, the air content may be near or at zero. If a drought persists for several months and the soil is depleted of available moisture, air increases as the water percentage diminishes.

PORE SPACE

Water and air are found in the interstitial spaces between soil particles. The supply of water, its rate of movement, and the availability of oxygen are all determined largely by the amount and size of these soil pores. In determining a given soil's pore space, the pedologist must take into consideration the *bulk density* and *particle density* of soil materials.

Bulk Density

The weight of an "oven dry" soil sample divided by its volume constitutes bulk density, or

$$\text{Bulk Density (BD)} = \frac{\text{Weight}}{\text{Volume}}$$

in either grams/cm^3 or pounds/ft^3. The greater the bulk density, the less the pore space. For example, solid rock material, with no pore space, will have a greater weight per volume, hence greater bulk density, than the same volume of shattered rock material of similar composition (Figure 3.1). The reason for the difference is the inability of rock fragments to fit together perfectly, thus creating unoccupied space between fragments and decreasing total weight per volume.

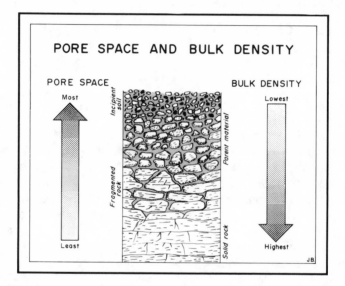

Figure 3.1 Pore space and bulk density vary inversely. As inorganic soil components decrease in size, the total pore space normally increases and bulk density decreases.

Particle Density

Unlike bulk density, which includes solids and pore spaces, particle density is the weight per unit volume of soil solids, as if the soil particles were compressed into a volume in which all the pore spaces were squeezed out.

$$\text{Particle Density (PD)} = \frac{\text{Weight of soil solids}}{\text{Volume of soil solids}}.$$

Note the difference between bulk density and particle density in the examples in Figure 3.2.

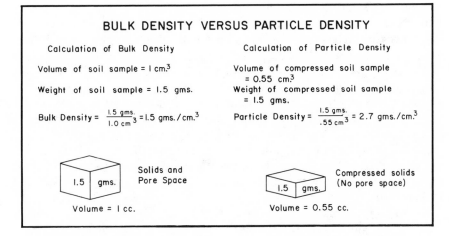

Figure 3.2   Examples of the calculation of bulk density and particle density.

Pore Space Size and Amount

Knowing the bulk and particle densities, it is possible to determine the percentage of pore space in mineral soils; that is, the portion of the soil occupied by either air or water. It is rather obvious that the total volume occupied by a given soil is the sum of pore space and solid space. If we know the value of either soil portion, we can determine the value of the other. The following equations are helpful:

$$\text{Percent Soil Solids} = \frac{\text{Bulk Density}}{\text{Particle Density}} \times 100, \text{ and}$$

$$\text{Percent Pore Space} = 100 - (\text{Percent Soil Solids}).$$

The sizes of pore spaces vary. Large particles produce large pores, and small particles such as clay create small pore spaces. Conversely, total pore space is greater and bulk density generally lower in fine-textured surface soils. This apparent paradox can be easily explained. Larger fragments permit the existence of larger individual spaces between particles than do the small fragments. Yet, the smaller fragments, although containing smaller spaces, have a greater total pore space due to the much greater number of particles in contact with one another per unit volume. As the finer fraction of the surface soil, such as clays, becomes dominant, individual pore spaces decrease further in size, and total pore space continues to increase (Figure 3.1). (An exception occurs in subsoils where interstitial spaces can often become clogged by colloid-sized materials and/or translocated clays, or when soils are compacted by machinery during cultivation.)

Pore space size is extremely important regarding moisture infiltration and storage, and soil aeration. Large pore spaces—*macropores*—result in rapid infiltration of moisture into the soil. This water, however, also will drain out of the large pores rapidly. Moisture storage thus tends to be very low. Such soils explain the occurrence of certain drought-resistant plant species on sandy soils in humid regions. Since plants depend primarily on water stored in the soil, meager moisture reserves result in stresses on vegetation during interprecipitation periods.

Small pore spaces—*micropores*—that are typical of clay soils restrict the infiltration of water into the soil. Water must move through the minute pore spaces very slowly. Likewise, drainage from the rooting zone is also very slow—indicating a higher water-holding capacity.

Since water drains out of large pores readily, macropores are usually filled with air and often referred to as *aeration pores*. Micropores are normally so small that the capillary films of moisture surrounding the soil particles saturate the pore space with water. Hence, they are commonly called *capillary pores*. Both the size of pores and the total amount of pore space are important characteristics of a soil.

SOIL MOISTURE

Precipitation results in water either on the surface of the land, within the soil itself, or at below-surface depths in the form of ground water. For atmospheric moisture to become part of the soil moisture reservoir, water must first infiltrate the surface layer of soil and percolate downward by gravitation and capillary forces.

Infiltration is the process by which water enters the soil surface and moves downward. The water will first replenish any lack of stored moisture. Thereafter, gravity moves any excess downward to become ground water. Each soil has a unique ability to infiltrate specific amounts of water over a given time period. This characteristic is called the soil's *infiltration capacity*. Clearly, rain may enter the soil at capacity rates only during or immediately following periods in which rainfall is in excess of the infiltration capacity.

Several factors operate simultaneously in affecting the infiltration capacity from one location to another or from one season to another. Some of the important elements are:

1. *Soil Moisture.* If the soil is dry at the beginning of a rain, the wetting of the top layer creates a strong capillary potential just below the surface, thus supplementing the gravitational force and increasing infiltration. On the other hand, when colloids—organic and inorganic—and clay particles are wetted, they may swell and reduce the infiltration capacity.

2. *Compaction Due to Rain.* Sandy soils are little affected by rain compaction. Fine-textured soils such as clays, however, are subject to the impact of raindrops, which compresses soil particles and reduces pore space and infiltration.

3. *Inwash of Fine Materials.* Fine particulate matter is held in suspension by rainwater on the soil surface. As infiltration begins, the soil acts as a filter, collecting the matter in pore spaces. As accumulation continues, interstitial spaces may become clogged and infiltration reduced.

4. *Topography.* Slope gradient, length, and shape are all significant in determining how rapidly water may run off, and, consequently, the length of time a given amount of water will remain at a certain point on the landscape.

Other factors are also important. Among them are: the viscosity (gluey consistency) of the infiltrating water, the proportion of air trapped in subsurface pore spaces, the degree of compaction and inwash reduction by vegetative cover, and the degree of compaction due to animals. Seasonal changes may occur due to changes in land use, characteristics of vegetation, cultivation practices, and temperature. These factors illustrate the great variability in infiltration capacity that can occur within short distances and through time.

Once water has infiltrated the soil surface, it moves into the main body of the soil, enveloping in a film of water each soil particle it contacts, filling micropores, and, if available in sufficient quantity, occupying the macropores as well. Only a portion of this water, however, is available for plant usage. Some of the water is held very tightly, resisting forces that attempt to displace it. This resistance to movement is known as *capillary tension* or *capillary potential* (a measure of the force required to remove this moisture). An examination of soil moisture indicates that several layers of water molecules are held at various levels of force. Understanding the ways in which water occurs in the soil provides a knowledge of moisture availability to plants and soil water movement.

Soil water is generally categorized as *hygroscopic, capillary,* or *gravitational (free).* The suction or tension that holds this moisture depends upon the amount of water that is present. The greater the amount, the less the tension (Figure 3.3). A soil need not be completely dry to restrict plant growth. Plant damage can occur when the work required to remove moisture from soil particles exceeds the plant's capabilities, even though water may account for well over 5 percent of the weight of the soil.

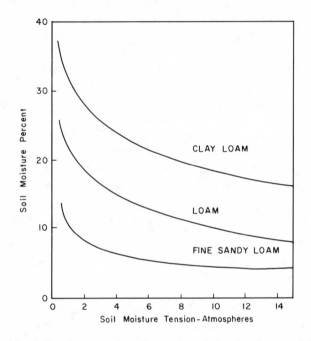

Figure 3.3   Soil moisture tension. Forces that retain water in the soil increase progressively faster as the soil dries. (*After Taylor,* The Yearbook of Agriculture: 1957. *Washington, D.C.: U.S. Government Printing Office, 1957, p. 64.)*

Hygroscopic Water

The film of water nearest to the soil particle is known as *hygroscopic water.* This form of water is held very rigidly to soil particles by *adhesion,* the attraction to solid surfaces for water molecules. Hygroscopic water is present in soils (including desert soils) under normal conditions, but due to its very firm attachment to the soil particle it is not available for plant use. This moisture can be removed from the soil, and its quantity measured, only by drying a soil sample in an oven under very high temperatures.

Capillary Water

Capillary water, also known as *water of cohesion*, envelopes the hygroscopic water that surrounds the soil particle. The attraction force in this case is the attraction of water molecules for each other. At the point of contact with the hygroscopic film, capillary water is held very firmly—although not as rigidly as the adhesion water—but as the film thickens moisture is held with less tension.

Plants obtain most of their water needs from the cohesion layer, even though not all of this water can be utilized. As the force required to remove moisture becomes too great (when the capillary film approaches the hygroscopic layer). the plant will show signs of moisture stress, and will wilt. This level of remaining moisture is known as the *wilting point* (Figure 3.4).

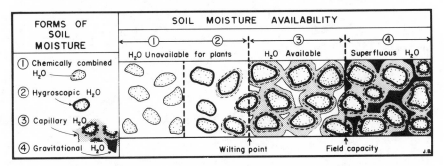

Figure 3.4 Forms of soil moisture. The amount of water available to plants is equal to the "field capacity" minus "wilting point."

Gravitational or Free Water

Following a prolonged rain or period of irrigation, the soil may have all interstitial spaces occupied by water; it is then considered *saturated*. Much of the soil moisture at this stage, however, is occupying macropores, which, having very little capillary potential, will rapidly drain once the external water supply ceases. Such water, held only temporarily within the soil, is referred to as *gravitational* or *free* water. Because it stays in the soil for only a very short period, its use by plants is quite limited. After gravitational drainage, the soil is said to be at *field capacity*.

Available Water

As we noted previously, the smaller the particle size, the greater are the number of particles and total pore space of the soil. With increased numbers of particles and surface area, the greater is the moisture-holding capacity of the soil—there are more particles to hold films of hygroscopic and capillary water, and fewer macropores from which water will escape. Hence, soil water storage is directly dependent upon solid particle size, structure, and organic matter content.

Available water may be considered as the difference between a soil's field capacity and the wilting point of plants (Figure 3.4). When fine particles dominate the soil the field capacity increases and the amount of water in hygroscopic form also increases. However, the proportion of available water to the total water held will decrease (Figure 3.5).

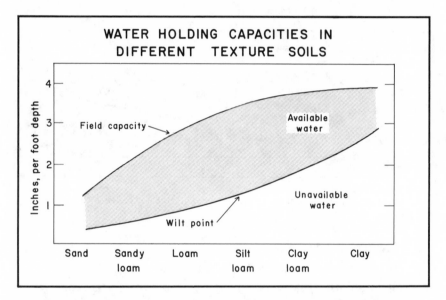

Figure 3.5   The above diagram illustrates the general relationship between soil moisture availability and soil texture.

The amount of available water in the rhizosphere (root zone of the soil) at any moment is largely dependent on regional climate patterns. Except for artificially adding water through irrigation, the primary source of soil moisture comes from naturally occurring precipitation, which infiltrates and replenishes the soil reservoir. Each climatic zone, in turn, places a demand for moisture upon the soil. This water requirement is induced by solar energy, which is capable either of converting free water in its liquid state to a gaseous form (evaporation), or converting water existing within the pore space of plants to a gaseous state (transpiration). The total capacity for such atmospheric moisture transformations is called *potential evapotranspiration* (PE); in other words, the amount of water that could be evaporated and/or transpired under conditions of optimal soil moisture and specific energy availability.

Varied precipitation and thermal zones over the earth's surface result in regional variations in moisture supply (precipitation) and demand (potential evapotranspiration). Thus, while vegetation in one region can survive periods between precipitation events—due to soil moisture extraction—without experiencing moisture stress, other areas are characterized by seasonal soil water depletions sufficiently severe to require physiognomic adaptations by plants.

A simple graphic method for describing the moisture "supply and demand" situation is shown in Figure 3.6 for two locations in North America. Brevard, North Carolina, has a low winter moisture requirement (less than ½″ per month) due to low temperatures and short daily periods of illumination. This need increases rapidly in the spring—as solar radiation becomes more intense and the period of illumination lengthens—reaching a peak of over 5″ in July, and then drops rapidly in the fall. In each month, precipitation exceeds the moisture demand. As soil moisture is utilized, it is replenished, and the excess precipitation either drains through the soil profile or is lost as surface runoff.

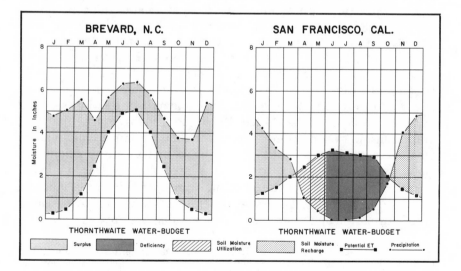

Figure 3.6 Water budgets: Brevard, North Carolina and San Francisco, California.

The moisture requirement of San Francisco, California, is higher in winter and lower in summer than that of Brevard. Although San Francisco's summer moisture demands are less, the ability of the atmosphere to supply precipitation is exceedingly low. Hence, transpiration and evaporation heavily tax the soil reservoir until the moisture supply is exhausted,[1] and a water deficiency exists.

Interrelationships between precipitation and potential evapotranspiration are vital aspects of regional pedogenic (soil formation) processes. Although we shall explore this topic in detail in later chapters, a few brief examples of the far-ranging effects of moisture supply and demand are appropriate here: (1) where precipitation exceeds potential evapotranspiration the chemical alteration of parent material is enhanced by frequent soil leaching; (2) the character of vegetation, hence the organic content of soil, is subject to the moisture supply and thermal regime; and (3) dehydration of soils during periods of moisture deficiency can lead to the precipitation and accumulation of minerals previously maintained in soil solutions.

Soil depth, texture, organic matter content, and structure—in combination—dictate the amount of water capable of being stored in a given soil. These soil components can also influence the rate of soil moisture release to the atmosphere, and more interestingly they themselves are products of the moisture-energy regime in which they play a role. A continuing short-term (weekly or monthly) computation of the moisture supply and demand relationships or *water-budget* (precipitation-potential evapotranspiration-soil moisture storage balance) of a site provides a current measure of the moisture available within the rhyzosphere. Such datum is beneficial, for instance, informing an agriculturist when to replenish soil moisture through irrigation before depletion has seriously harmed crops.

Figures 3.7 and 3.8 show the available moisture and its seasonal variation by area in the United States. Both maps use the potential evapotranspiration measure. Remember that it is impossible to know whether a climate is moist or dry by knowing only precipitation amounts. It is necessary to compare precipitation with potential evapotranspiration. Likewise, such moisture-energy interrelationships must also be considered to understand the effect of climatic variables in soil formation and upon the moisture available to plants.

---

[1] The rate of moisture release from the soil is dependent upon the quantity of available water in the soil reservoir. At field capacity, release approximates demand. As the soil dries out, moisture release is a function of the percentage of remaining soil moisture and energy demands, that is, at 20 percent field capacity, release may approximate 20 percent of demand.

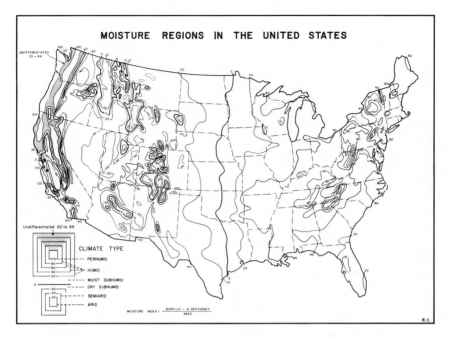

Figure 3.7 Moisture regions in the United States (*after C. W. Thornthwaite*).

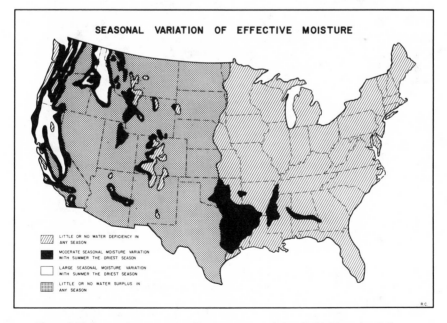

Figure 3.8 Seasonal variation in effective moisture (*after C. W. Thornthwaite*).

SOIL ATMOSPHERE

Soil air exists in the pore spaces that are not occupied by water, and its composition of gases is unique. Unlike the free atmosphere, soil atmosphere is rich in carbon dioxide (10 to 100 percent greater than above the surface) and has lower quantities of oxygen. The reason is that plant roots and the organisms living in the soil remove oxygen from, and respire carbon dioxide into, the interstitial spaces. Most crops cannot grow if the carbon dioxide content in the root zone is too high or the oxygen too low. A constant exchange between the soil and free atmosphere must take place, the free atmosphere supplying oxygen and the soil diffusing carbon dioxide into the free atmosphere. For soils to be well aerated, there must be sufficient open interstitial spaces to permit essential gases to move easily in and out of them. Saturated soils obviously have little space available for air.

Site and Time Affecting Soil Characteristics

4

The major components of soil—inorganic material, organic matter, water, and air—have been individually considered. It is obvious that the dominance of one component over the other can produce soils of varied characteristics. Yet the question remains, Why do soils occurring on similar parent material have different chracteristics in one location from those adjacent to or distant from them? One obvious factor is vegetation. In Chapter II it was shown that organic content and coloration of soil were strongly associated with the type of vegetative cover. Yet, there are other aspects of pedogenesis, or soil formation, that also are significant in producing variations. The most important are topography, climate, and the length of time soil formation has been occurring.

TOPOGRAPHIC POSITION

Topographic position is an important factor in determining the characteristics of a given soil. Landscapes in a mountainous region may exist in a variety of climatic and vegetative zones due to vertical changes in both temperature and moisture. As a consequence, exposed parent materials of similar composition within relatively short distances of one another may be subject to change by different sets of soil-forming processes. Furthermore, local relief and drainage characteristics have an affect upon soil properties even within a uniform climatic regime.

Elevation Differences

Steep mountain slopes often speed up the removal of loose surface material by enhancing the erosive power of gravity and running water, not allowing deep soil to develop. Buol[1] describes the soils of mountain slopes in Arizona as follows:

> These soils result largely from the interaction of topography and time. The steepness of the slopes allows for the rapid removal of friable soil material by water and gravity. Therefore, the soil surface is constantly lowered at such a rapid rate that many of the soil-forming processes controlled by climate and organisms do not have time to impart a significant influence on the soil development. Instead, most of the soil characteristics—namely, texture and chemical composition—are closely related to those of the parent rock.
>
> The influences of climate and vegetation are observed mainly in the color and organic content of the surface layer of the soil. The cooler and wetter areas, which usually support more vegetation, have a higher organic matter content, and thus a darker color in the surface layer, than where the climate is hotter and drier. Organic matter is also oxidized at a slower rate in a cold climate than in a warm one.
>
> Being shallow over rock material, these soils store little water to be used by growing plants; consequently, unless rains are frequent vegetation is sparse.

Upthrust mountain masses exist in a variety of climatic regions due to temperature lapse rates and orographic (mountain-induced) precipitation. Figure 4.1 shows the relationship between average annual temperature and precipitation with increasing height in the Catalina Mountains near Tucson, Arizona. The pattern of decreasing temperature and increasing precipitation with increased elevation produces a series of climatic environments. Associated with this pattern in Arizona is a variety of floral habitats, ranging from the scrub of the hot subtropical desert to the fir forests of the humid and cool high-elevation zone. A similar relationship is shown in Figure 4.2 for the central Sierra Nevada.

The result of such varied vegetation and climates interacting with the parent material is a succession of soil types with differing features. An examination of soil characteristics in the mountains of the southwestern United States revealed the following:[2]

---

[1] S.W. Buol, *Soils of Arizona* (Tucson: Agricultural Experiment Station, University of Arizona, 1966), p. 3.

[2] W. P. Martin and Joel E. Fletcher, "Vertical Zonation of Great Soil Groups on Mt. Graham, Arizona, as Correlated with Climate, Vegetation, and Profile Characteristics," *Technical Bulletin* #99 (Tucson: College of Agriculture, University of Arizona, 1943), pp. 144-47.

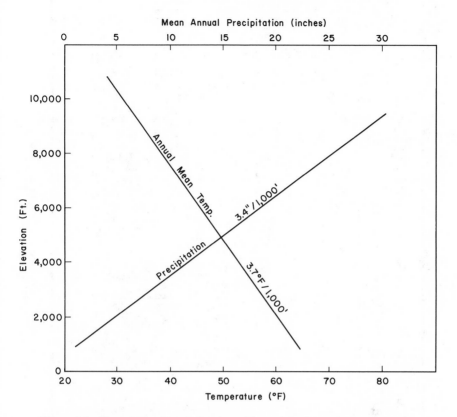

Figure 4.1 The relationship of elevation with precipitation and temperature in the Catalina Mountains of Southern Arizona.

1. Organic content of the soils increased with elevation.
2. Total porosity and water-holding capacity—associated with organic matter—increased with elevation.
3. pH values decreased with increased elevation; desert soils were alkaline and had calcareous (containing calcium or a calcium compound) layers.
4. C-N ratio increased with elevation.

Local Relief

Soils with identical parent material, existing within a uniform climatic region, and affected by parallel soil-forming factors still may have varied features due to local relief and/or internal drainage characteristics. The name for such a group of related soils is *catena,*

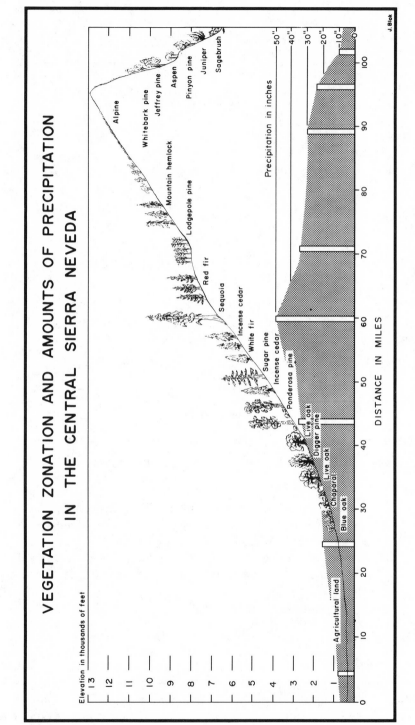

VEGETATION ZONATION AND AMOUNTS OF PRECIPITATION IN THE CENTRAL SIERRA NEVEDA

Figure 4.2 Relationship between elevation, precipitation, and vegetation.

meaning "chain."[3] Figure 4.3 shows a valley slope profile in the Machkund Basin of India. Soil boundaries are marked and each soil classified by number. Table 4 gives a description of each soil's characteristics from which the following observations have been made regarding the catena:

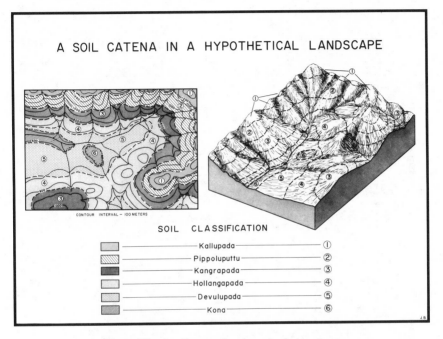

Figure 4.3   A soil catena in a hypothetical landscape.

The soils on the hilltops and the slope of hillsides, namely, soils of the Kallupada series, are shallow, occasionally attaining moderate depth. As the slope eases up to the foothill areas, the soils attain greater depth. The soils of the valley and depressions are very deep. The data of mechanical and chemical composition of the soils . . . show that there has been an increase in the content of clay in the soils from hilltop to valley.

From the foregoing discussion, it would be apparent that there has been lateral translocation of finer materials as also of soluble constituents on account of lateral movement of water, which flows down as runoff. The above situation is suggestive of lack of deep percolation through depth in hilltops and slopes,

---

[3] A *toposequence* is also a small soil association occurring in the same climatic zone, under similar vegetation, and exhibiting differences related to topography. However, these associated soils do not have to possess a common parent material.

Table 4   Physical and Physico-chemical Properties of Horizon Soil Samples of the Recognized Soil Series in the Machkund Basin.

No.	Soil series and depth (em)	Mechanical constituents (expressed as percentage on air-dry basis)			Physico-chemical properties				
		Sand	Silt	Clay	Total c.e.c. m.e. 100 g	T.E.B. m.e. 100 g	Base satura-tion (%)	pH	B.E.C. m.e.% calcu-lated on clay
1	Kallupada								
	0-10	71.4	9.0	19.9	6.7	3.2	47.5	5.4	33.8
	10-35	54.3	18.5	27.1	8.8	4.8	54.5	5.6	32.4
	35-45	45.8	21.3	32.8	9.2	4.8	52.1	5.6	28.1
2	Pippoluputtu								
	0-12.5	73.9	10.3	15.8	7.2	3.8	52.7	5.4	45.4
	12.5-40	61.6	13.7	24.7	8.4	4.0	50.9	5.5	34.0
	40-75	48.4	14.2	37.4	10.8	5.5	51.1	5.7	29.0
3	Kangrapada								
	8-15	55.8	21.4	22.7	6.8	3.5	51.8	5.6	30.0
	15-42	49.9	22.6	27.5	8.9	4.5	50.9	5.6	32.5
	42-90	44.5	22.9	32.6	10.6	5.5	51.7	5.7	32.7
4	Hollangapada								
	0-15	56.7	21.1	22.1	8.4	4.3	51.4	5.4	38.0
	15-57.5	58.7	17.2	24.0	9.3	5.1	54.4	5.4	38.7
	57.5-150	58.9	15.6	25.4	10.8	5.7	52.7	5.6	42.4
5	Devulupada								
	0-12.5	53.9	19.5	26.6	8.8	5.2	59.0	5.4	33.1
	12.5-45	49.8	20.8	29.4	13.2	7.9	60.1	5.4	44.9
	45-85	43.9	21.8	34.2	22.9	13.3	58.1	5.6	66.9
	85-150	35.6	23.6	40.7	26.3	15.7	59.9	5.6	64.3

Source: S. V. Govinda Rajan and N. R. Datta Biswas, "Development of Certain Soils in the Subtropical Humid Zone in Southeastern parts of India, Genesis and Classification of Soils of Machkund Basin," Soils and Tropical Weathering (Paris: UNESCO, 1971), p. 83.

giving rise to shallow soil. Further accelerated erosion due to slope conditions depletes the soil materials from convex and steep-slope situations, giving rise to shallow soils. Deep percolation of water in the other areas accentuating weathering and receipt of soil material from the upland and through the drainage courses have given rise to deep soils in valleys and depressions.

The cation exchange capacity of the soils also varies, being lower in respect of soils of the hills and hillslopes and higher in respect of the soils in the low-lying situation. The high content of dibasic constituents in the low-lying soils coupled with high moisture regime appears to have helped resilification of the secondary clay minerals enriching the soils in contents of 2:1 lattice minerals of high-exchange capacity. The derived values of cation exchange capacity definitely show a transition from hills to low-lying areas.[4]

THE SOIL PROFILE: A FACTOR OF TIME

Soil is a naturally occurring body undergoing continual change. Its properties are the combined effects of *climate* and *biotic activity* (plants and animals) acting upon *parent material*, as conditioned by *topography* over periods of *time.*

When soil scientists analyze a given soil in its natural setting, they concern themselves with the variation in the soil's composition and character with depth; that is, they examine a soil profile, which is a vertical cross section of a *pedon.* These two terms may often become confused. Following is the distinction between them:

A pedon is a three-dimensional body of soil with a lateral area large enough to permit the study of horizon shapes and relations (Figure 4.4). Its area contains a minimum of 1 square meter (approximately 1 square yard). If the soil horizons are laterally discontinuous, however, the area needed to define the pedon may increase to as much as 10 square meters (10 square yards).

A soil profile is a vertical two-dimensional slice through the pedon. It extends from the surface of the earth through the altered layers and into the nonsoil or parent material below (Figure 4.5).

A description of the horizon properties as they vary with depth provides information regarding plant root environments, nutrient and moisture availability, soil susceptibility to surface erosion and internal leaching, clay minerology, and other data upon which wise land-use planning may be based. Each *soil individual*[5] is the synthesized product of all soil-forming factors operating *in situ*[6] and has unique characteristics associated with its particular stage of development.

[4] S. V. Govinda Rajan and N. R. Datta Biswas, "Development of Certain Soils in the Subtropical Humid Zone in Southeastern Parts of India, Genesis and Classification of Soils of Machkund Basin," *Soils and Tropical Weathering* (Paris: UNESCO, 1971), p. 80.

[5] A soil individual is a soil distinct in characteristics from adjacent soil bodies or nonsoil materials.

[6] *In situ* (Latin) meaning in its original place.

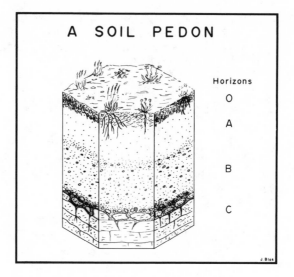

Figure 4.4   The soil pedon in its three-dimensional form.

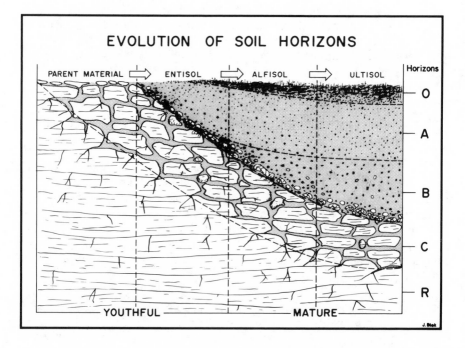

Figure 4.5   A sequence of soil profiles from youthful through mature.

Exposed crustal materials are subject to weathering processes that produce physically and chemically altered inorganic debris. Such material—given sufficient time and favorable atmospheric conditions—can have established upon and within it a community of plants and animals. The accumulation of their organic residues provides a new complex between the earth's crust and the atmosphere—a body of material consisting of both organic and inorganic matter. This complex is conducive to an expanded biological community. Microorganisms, including bacteria, protozoa, and a host of other flora and fauna, become increasingly more abundant—feeding on organic remains and releasing nutrients for further plant growth. Examination of the soil at this state in development reveals a surface layer that is darkened in color because of organic matter in various states of decay overlying the parent material.

If soil-forming processes continue and the rate of soil loss by erosion is less than the rate of rock weathering, the soil will continue to deepen. As deepening takes place, the soil will also vary in characteristics with depth. The surface layer (topsoil) of the mineral soil, under the influence of percolating water, will have *eluviated* (removed) from it soluble minerals and suspended colloidal-sized particles. The colloidal particles—clay, organic matter, and oxides of aluminum and iron—move only a few feet at most before they become lodged, creating an *illuvial* (accumulation) layer. The presence of both an eluvial and an illuvial layer, called the soil's A and B horizons respectively, indicate an advanced stage of soil evolution. That is, if a soil contains only an A horizon resting upon partially altered parent material (the soil's C horizon), it is said to be young, or *immature*. There has not been enough in the given climate, along with related soil-forming processes, to produce horizon differentiation. Soil characteristics at this stage are organic matter accumulation and properties strongly influenced by the parent material. A soil attains a *mature* stage of development with the formation of the illuvial B horizon (Figure 4.5). The A and B horizons together are known as the *solum*, or true soil.

Soil Horizons

The soil horizons are layers of soil or soil material lying approximately parallel to the land surface. They differ from adjacent genetically related layers in physical, chemical, and biological properties or characteristics, such as color, structure, texture, consistency, kinds and numbers of organisms, and degree of acidity or alkalinity. Brief descriptions of the major soil horizons follow:

Horizon Designation	Description
O	Organic horizons of minerals soils. Horizons: (1) formed or forming in the upper part of mineral soils above the mineral part; (2) dominated by fresh or partly decomposed organic material; and (3) containing over 30 percent organic matter if the mineral fraction is greater than 50 percent clay, or over 20 percent organic matter if the mineral fraction has no clay. Intermediate clay content requires proportional organic matter content.
A	Mineral horizons consisting of: (1) horizons of organic matter accumulation formed or forming at or adjacent to the surface; (2) horizons that have lost clay, iron, or aluminum with resultant concentration of quartz or other resistant minerals of sand or silt size; or (3) horizons dominated by (1) or (2) above but transitional to an underlying B or C.
B	Horizons in which the dominant feature or features is one or more of the following: (1) an illuvial concentration of silicate clay, iron, aluminum, or humus, alone or in combination; (2) a residual concentration of sesquioxides or silicate clays, alone or mixed, that has formed by means other than solution and removal of carbonates or more soluble salts; (3) coatings of sesquioxides adequate to give conspicuously darker, stronger, or redder colors than overlying and underlying horizons in the same sequum[7] but without apparent illuviation of iron and not genetically related to B horizons that meet requirements of (1) or (2) in the same sequum; or (4) an alteration of material from its original condition in sequums lacking conditions defined in (1), (2), and (3) that obliterates original rock structure, that forms silicate clays, liberates oxides, or both, and that forms granular, blocky, or prismatic structure if textures are such that volume changes accompany changes in moisture.
C	A mineral horizon or layer, excluding bedrock that is either like or unlike the material from which the solum is presumed to have formed, relatively little affected by pedogenic processes, and lacking properties diagnostic of A or B but including materials modified by: (1) weathering outside the zone of major biological activity; (2) reversible cementation, development of brittleness, development of high bulk density, and other properties characteristic of fragipans; (3) gleying; (4) accumulation of calcium or magnesium carbonate or more soluble salts; (5) cementation by accumulation of calcium or magnesium carbonate or more soluble salts; or (6) cementation by alkali-soluble siliceous material or by iron and silica.
R	Underlying consolidated bedrock, such as granite, sandstone, or limestone.

[7] *Sequum* is a sequence of an eluvial horizon and its subjacent B horizon, if present.

The major horizons are usually subdivided. One B horizon, for example, might be subdivided into a B1, B2, and B3 series of layers: B1 being more like the B rather than the A horizon, but transitional to the A horizon; B2 having the dominant characteristics of the B horizon; and B3 being more like the B rather than the C horizon, but transitional in nature to the C horizon.

An extremely complex soil may be described in even greater detail. A B2 horizon can be subdivided into B21, B22, and B23 horizons. Figure 4.6 illustrates some possible horizon subdivisions.

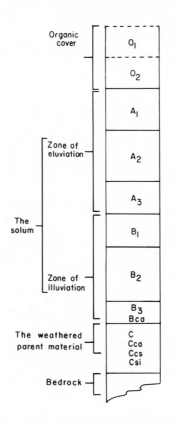

Figure 4.6 An example of possible horizon subdivisions.

The concept of the mature soil with horizon differentiation is extremely important to the understanding of the global variation of regionally distributed soils. Parent material, topography, and time may all be viewed as passive factors in soil development, whereas climate and vegetation are active agents. A strong regional relationship exists between climate and vegetation, for example, humid regions-forests, semiarid areas-grasslands, and arid climates-deserts. Hence, it is expected that pedogenetic processes similarly occur on a macroscale over broad regions of the earth, producing mature soils that have similar traits throughout. This world pattern of mature soils is discussed in Chapters 5 through 12.

Soil Classification

5

Soil characteristics vary through time and space. A complete analysis of all the constituents of several soil profiles from a farmer's single field, in fact, would reveal that each pedon is unique—no two soils are exactly alike. Just as there are some (however small) differences in even the most seemingly perfectly matched identical twins, pedons also vary over the earth's surface. On the other hand, those soils that are affected by similar environmental controls do tend to exhibit common properties related to their relatively uniform genetic background, at the same time showing differences derived from being in a particular niche in a particular region.

The classification of soils is designed to satisfy practical needs. In a 1938 classical treatise on soil classification, Baldwin, Kellogg, and Thorp state: "Man has a passion for classifying everything. There is reason for this; the world is so complex that we could not understand it at all unless we classified like things together. Just as plants, insects, birds, minerals, and thousands of other things are classified, so are soils."[1]

In relation to soils, the "passion" to classify is not a recent phenomenon associated with modern technology and mechanized cultivation, but has probably been a conscious factor since the

[1] Mark Baldwin, Charles E. Kellogg, and James Thorp, "Soil Classification," *1938 Yearbook of Agriculture: Soils and Man* (Washington, D.C.: U.S. Govt. Ptg. Office, 1938), p. 979.

beginning of agriculture. There is evidence that the Chinese recognized different kinds of soils, and named them, over 4,000 years ago. This early classification was made largely on soil color and structure. Within the last 150 years or so, data from soil analyses have become more abundant and have resulted in more sophisticated soil classification methods. Seeking the causes behind the recurrence of regionally distributed soil features, investigators in the early and mid-nineteenth century attempted to categorize soils on the basis of parent material—they assumed that the parent material was the primary factor in determining soil characteristics. This approach applied well to young soils, but provided unsatisfactory results on many mature sites where climate and vegetation had altered the parent material. In the latter half of the nineteenth century a fresh approach to pedology was initiated by Russian scientist V.V. Dokuchaiev. Soil classification now focused on the concept of soil as an independent natural body exhibiting distinct characteristics that resulted from the interactions of climate, parent material, flora and fauna, geomorphic factors, and time.

The Dokuchaiev school had a profound impact on future pedologic research, and the publications of his first students mark the inception of modern soil science. No longer was soil viewed simply as being produced by the physical and chemical alteration of its parent material. Rather, it was considered to be the evolutionary product of a dynamic system: (1) whose current status was the result of several energy imputs; (2) that varied its characteristics through time; (3) whose developmental factors could be identified; and (4) that, due to the varying degree of dominance of specific pedogenetic agents, had recognizable variations that could be shown on maps.

This new school of soil science was founded in Russia about 1870 under Dukuchaiev's leadership. Research focused on determining general soil characteristics and their areal distribution. Several genetic soil classification plans that came from those early investigations stressed the interrelated role of climate and vegetation as the primary pedogenetic agent.

Other countries had very little knowledge of the Russian advances in soil investigation, however, until the 1914 publication in Berlin of K.D. Glinka's *Die Typen der Bodenbildung, ihre Klassifikation und Geographische Verbreitung.* (Glinka was a student of Dokuchaiev's.) Then, C. F. Marbut became familiar with Glinka's writings, and through his influence as chief of the U.S. Soil Survey soil identification and mapping in the United States turned from a geologic orientation to one "based primarily on soil profile studies and their genetic implications."[2] Aligning with the Dokuchaiev philosophy, Marbut developed a genetic soil classification system that utilized considerable Russian terminology (Table 5).[3]

Table 5   Soil Categories by Marbut: 1938

Category VI	Pedalfers (VI-1)	Pedocals (VI-2)
Category V	Soils from mechanically comminuted materials. Soils from siallitic decomposition products. Soils from allitic decomposition products.	Soils from mechanically comminuted materials.
Category IV	Tundra. Podzols. Gray-brown Podzolic soils. Red soils. Yellow soils. Prairie soils. Lateritic soils. Laterite soils.	Chernozems. Dark-brown soils. Brown soils. Gray soils. Pedocalic soils of Arctic and tropical regions.
Category III	Groups of mature but related soil series. Swamp soils. Glei soils. Rendzinas. Alluvial soils. Immature soils on slopes. Salty soils. Alkali soils. Peat soils.	Groups of mature but related soil series. Swamp soils. Glei soils. Rendzinas. Alluvial soils. Immature soils on slopes. Salty soils. Alkali soils. Peat soils.
Category II	Soil series.	Soil series.
Category I	Soil units, or types.	Soil units, or types.

Marbut stressed the distinction between the dynamic (climate and biologic) and the passive (parent material, topographic position, and time) soil development factors. He also distinguished the soil proper (the solum) from its underlying geologic material (C horizon) and recognized the geographic expression of soils in the emphasis he placed on mature (zonal) soils—category IV in his classification system (Figure 5.1). These were considered to be "great soil groups"

[2] Charles E. Kellogg, "Why a New System of Soil Classification?" *Soil Science*, LXXXXVI, no. I (July 1963), 3.

[3] Marbut proposed an initial system of soil classification in 1927 that gave strong emphasis to mature soils. He continued to refine the system until it progressed to his 1936 version, which is provided in Table 5.

and consisted of hundreds of soil series that differed from one another due to variations in parent material, relief, and age, yet all exhibited the same general sort of profile. The definition of the soil series placed considerable emphasis on properties that were important to soil behavior. "Ranges in properties of series were narrow in highly productive soils, and wider in soils of low productivity or unsuited to farming."[4]

SOIL GROUPS OF THE UNITED STATES

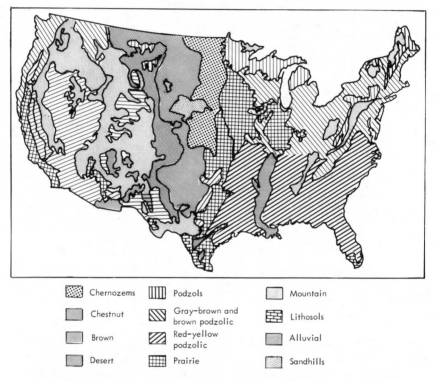

	Chernozems		Podzols		Mountain
	Chestnut		Gray-brown and brown podzolic		Lithosols
	Brown		Red-yellow podzolic		Alluvial
	Desert		Prairie		Sandhills

Figure 5.1 Great soil groups of the United States according to the Marbut classification scheme.

A demand for a major change in classification criteria came about during World War II when the great need for food and fiber motivated worldwide government-sponsored soil research and resulted in the identification of numerous soils previously unclassified. Immediately after the war serious attempts were made to incorporate

---

[4] Guy D. Smith, "Lectures on Soil Classification," *Pedologie*, IV (Ghent, Belgium: State University of Ghent, 1965), 123.

all of the known soil series, including those newly discovered, into the Marbut Classification Scheme. The effort failed even after the attempt to revise the definitions of the great soil groups (Table 6). The well-known soil scientist, C. E. Kellogg, remarked that most systems of soil classification in general use as late as 1950 had serious faults. Several emphasized pedogenesis and required assumptions regarding the soil's character under virgin conditions. Soils that had been severely eroded, drastically reworked, or made from transported materials were difficult to place within such genetic frameworks. A new classification method seemed necessary, one that would avoid past weaknesses and in which similar soils, whether cultivated or virgin, would be placed within the same taxon (group or entity).[5]

UNITED STATES COMPREHENSIVE SOIL CLASSIFICATION SYSTEM

During the 1950s the Soil Survey Staff of the U.S. Department of Agriculture, under the chairmanship of Guy D. Smith, assumed the responsibility for developing the United States Comprehensive Soil Classification System as a direct response to the recognized weaknesses of methods currently in use. The comprehensive methodology was completely new above the level of the soil series and was developed by a successive set of *approximations*, each of which was circulated to professional soil scientists for test applications and criticisms. These approximations, with successive supplements and revisions, became the body of data from which the current Comprehensive Classification System was derived.

The taxa[6] above the soil series level were given completely new names, because much of the terminology previously used to describe soil characteristics had been redefined many times and had held different meanings at different periods and in different parts of the world. Although the focus of the present system is based on the properties of the soil as it exists today (eliminating the shortcomings of previous systems that identified soil on the basis of the properties it would possess under virgin conditions), it does not completely ignore genesis. Since soil properties are directly related to soil development, genesis is implicitly considered in the current methodology.

[5] Kellogg, "Why a New System of Soil Classification?" p. 3.

[6] Individuals or items in any type of population may be grouped according to common characteristics or properties. A group so formed is known as a class or taxon (plural taxa).

Table 6   Soil Classification in the Higher Categories (Revised: 1949)

Order	Suborder	Great Soil Groups
Zonal soils	1. Soils of the cold zone.	Tundra soils
	2. Light-colored soils of arid regions.	Desert soils. Red Desert soils. Sierezem. Brown soils. Reddish-Brown soils.
	3. Dark-colored soils of semi-arid, subhumid, and humid grasslands.	Chestnut soils. Reddish Chestnut soils. Chernozen soils. Prairie soils. Reddish Prairie soils.
	4. Soils of the forest-grass-land transition.	Degraded Chernozem. Noncalcic Brown or Shantung Brown soils.
	5. Light-colored podzo-lized soils of the tim-bered regions.	Podzol soils. Gray wooded, or Gray Podzolic soils. Brown Podzolic soils. Gray-Brown Podzolic soils. Red-Yellow Podzolic soils.
	6. Lateritic soils of forested warm-temper-ate and tropical regions.	Reddish-Brown Lateritic soils. Yellowish-Brown Lateritic soils. Laterite soils.
Intrazonal soils	1. Halomorphic (saline and alkali) soils of imperfectly drained arid regions and littoral deposits.	Solonchak, or Saline soils. Solonetz soils. Soloth soils.
	2. Hydromorphic soils of marshes, swamps, seep areas, and flats.	Humic Gley soils (includes Wiesenboden). Alpine Meadow soils. Bog soils. Half-Bog soils. Low-Humic Gley soils. Planosols. Ground-Water Podzol soils. Ground-Water Laterite soils.
	3. Calcimorphic soils	Brown Forest soils (Braunerde). Rendzina soils.
Azonal soils.		Lithosols. Regosols (includes Dry Sands). Alluvial soils.

Six levels of generalization have been established, and soils are placed in a common taxon only if there is evidence of one or more dominant processes sufficiently affecting the soil to produce common diagnostic horizons or features.[7] At the order level, only a few similarities are present. In the lowest category (soil series), there is relatively complete homogeneity in both soil features and their genesis. The criteria for separation of the various taxa is briefly summarized below:

I. Order Level—A given soil must possess common properties indicating similarity in kind and strength of pedogenic processes, such as the presence or absence of major diagnostic horizons.

II. Suborder Level—This class has genetic homogeneity. Subdivision of orders are according to the presence or absence of properties associated with wetness, soil moisture regimes, major parent material, and vegetational effects as indicated by key properties.

III. Great Group Level—Differentiation is based upon similar kind, arrangement, and degree of expression of horizons, with emphasis on the upper sequum; base status; soil temperature and moisture regimes; and the presence or absence of diagnostic layers (plinthite, fragipan, and others).

IV. Subgroup Level—Provides for the central concept taxa of the great group and for properties indicating intergradations to other great groups, suborders, and orders, and to extragradation to "not soil."

V. Family Level—This class identifies features of importance to plant root growth, such as broad textural characteristics, mineralogical composition, and soil temperature and reaction.

VI. Series Level—Soils are separated on the kind and arrangement of horizons; color, texture, structure, consistency, and reaction of horizons; and the chemical and mineralogical properties of the horizons.

The order represents the greatest degree of soil generalization and consists of ten categories, each possessing an identifying name ending in *sol* (from L. *solum*—soil). They are listed in Table 7 along with their root word derivations, percentage of areal dominance, and approximate equivalents in the latest revised Marbut System (1949).

---

[7] Each soil exhibits distinct features that are characteristic of the environment in which it has formed. These features are said to be "diagnostic" in that they distinguish the soils in one taxon from those in another.

Table 7 Soil Orders Name Derivation, Areal Significance, and Marbut Equivalents.

Soil Order	Derivation of Root Word	% of Total World Soils†	Rank (Total Area)	Marbut Equivalents
Entisols	Recent.	12.5	4	Azonal soils, some Low Humic Gley.
Vertisols	L: *verto* = to turn.	2.1	9	Grumusols.
Inceptisols	L: *inceptum* = inception, beginning.	15.8	2	Ando, Sol Brun Acide, some Brown Forest, Low Humic Gley, Humic Gley.
Aridisols	L: *aridus* = dry.	19.2	1	Desert, Reddish Desert, Serozem, Solonchak, some Brown and Reddish Brown soils, associated Solonetz.
Mollisols	L: *mollis* = soft.	9.0	6	Chestnut, Chernozem, Brunizem, Rendzina, some Brown, Brown Forest, associated Humic Gley, and Solonetz.
Spodosols	Gr: *spodos* = wood ash.	5.4	8	Podzols, Brown Podzolic, Groundwater Podzols.
Alfisols	Alf: combined from aluminum and iron.	14.7	3	Gray-Brown Podzolic, Gray Wooded, Noncalcic Brown, Degraded Chernozem, associated Planosols and Half-Bog.
Ultisols	L: *ultimos* = ultimate.	8.5	7	Red-Yellow Podzolic, Reddish-Brown Lateritic, associated Planosols, and some Half-Bogs.
Oxisols	Oxi: from oxide	9.2	5	Laterite soils, Latosols.
Histosols	Gr: *histos* = tissue	.8	10	Bog soils.

†An additional 2.8 percent of the world total includes ice fields, unclassified lands, and others.

Each of the mineral soils at the order level occupies a position within a hierarchy that is largely based upon *degree of weathering*, for example, degree of mineral alteration and profile development (Figure 5.2). Their global distribution is shown in Figure 5.3. The *Entisols* are youthful soils characterized as having no natural genetic horizons or at best only the beginnings of such. *Vertisols* are soils containing large amounts of montmorillinite (expanding) clay. Because of alternate shrinking and swelling this soil will become "inverted" with time. *Inceptisols* are soils that are beginning to show development of genetic horizons. They lack evidence of extreme weathering and are not sufficiently developed to be classified in one of the remaining seven orders. *Aridisols* are light-colored soils of desert climatic regimes. *Mollisols* are typically found in grassland environments. They have a soft, thick, dark-colored surface horizon. *Spodosols* are primarily found in cool and humid forested regions. They contain an illuvial subsurface horizon in which amorphous organic matter and aluminum with or without iron have accumulated. *Alfisols* are strongly weathered soils in which

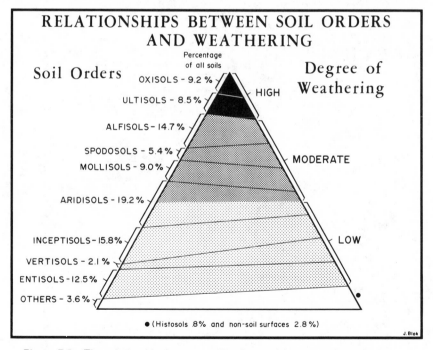

Figure 5.2 The relationship between soil orders and intensity of weathering. (The percentages provide the approximate areal extent of each Soil Order.)

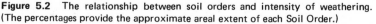

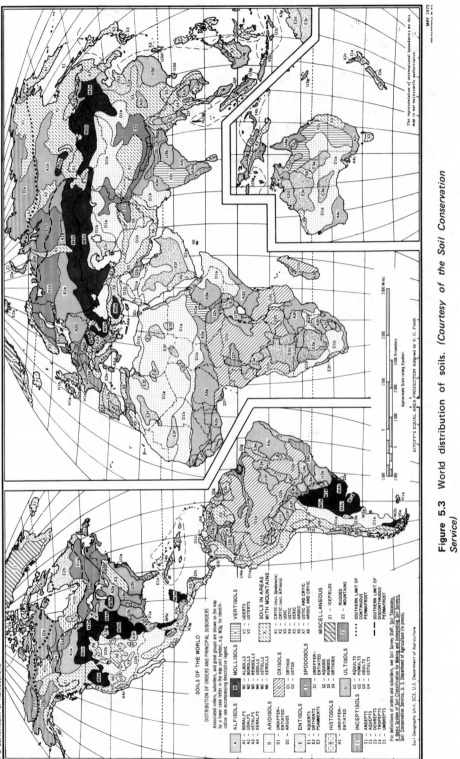

Figure 5.3 World distribution of soils. *(Courtesy of the Soil Conservation Service)*

translocation of aluminum and iron has taken place; yet they retain a relatively high base saturation. The *Ultisols* are extremely weathered soils with very low retention of bases. *Oxisols* are found within the tropics. They contain subsurface horizons consisting of a mixture of hydrated oxides of iron and/or aluminum and 1:1 lattice clays. The *Histosols* represent the organic soils and are composed primarily of vegetative debris in various stages of decomposition.

As a result of macro-scale pedogenesis, there is regional homogeneity in diagnostic soil characteristics. A close examination of Figure 5.3 reveals a logical pattern to the distribution of soil orders that can be readily associated with relatively extensive climatic and vegetative realms. Yet each order is made up of hundreds of subtypes, each of which has attributes associated with its unique intraregional environment. The taxa of the lower categories provide an orderly identification of such differences.

The orders are differentiated into suborders on the basis of chemical and/or physical properties that indicate drainage conditions or that possess genetic differences due to climate and vegetation. The name of the suborder is comprised of two syllables. The last syllable always indicates the order, for example, *Ult* for Ultisol. The prefix identifies a characteristic unique to the particular suborder, such as *aqu* for aqua (L: *aqua*, water), meaning wet. Hence, an Ultisol exhibiting signs of wetness would be classified in the suborder *Aquult*. Likewise, an Entisol with a sandy surface horizon would be considered a Psamment (Gr: *psammos*, sand). The formative elements identifying factors of importance at the suborder level are shown in Table 8.

The suborders are broken down into *great groups* on the basis of the kind and array of diagnostic horizons. The name of the great group indicates the type of features present by prefixing one or more formative elements onto the correct suborder name. For example, a *Plinthaquult* is a wet Ultisol that has plinthite within the profile, and a *Quartzipsamment* is a sandy Entisol comprised primarily of quartz crystals. (See Table 9 for the formative elements for names of the great groups.)

The great groups are divided into subgroup categories that indicate to what extent the central concept of the great group is expressed. A *typic Plinthaquult*, for example, is a soil typical of the great group Plinthaquult.

The lowest taxonomic categories are family and series. The families stress features of importance to plant growth, such as texture, mineralogy, reaction class, and temperature. The series identify the individual soil, named after a natural feature or place where the soil was first discovered. A complete soil classification for the Pantego series, along with its profile description, follows.

PANTEGO SERIES[8]

The Pantego series is a member of the fine-loamy, siliceous, thermic family of Umbric Paleaquults. These soils have black or very dark gray, fine, sandy, loam A horizons and grayish, sandy, clay loam Bt horizons.

Horizon	Depth (Inches)	Typifying Pedon: Pantego fine sandy loam—Cultivated field. (Colors are for moist soil.)
Ap	0-10	Black (10 YR 2/1) fine sandy loam; weak fine granular structure; very friable; many fine roots; very strongly acid; gradual wavy boundary (6 to 12 inches thick).
A12	10-18	Very dark gray (10 YR 3/1) fine sandy loam; weak fine granular structure; friable; very strongly acid; clear smooth boundary (4 to 8 inches thick).
B21tg	18-27	Very dark gray (10YR 3/1) sandy clay loam; weak fine subangular blocky structure; friable; patchy clay films on ped faces and in pores; very strongly acid; gradual wavy boundary (9 to 18 inches thick).
B22tg	27-42	Gray (10YR 5/1) sandy clay loam; few fine and medium distinct mottles of brownish yellow (10YR 6/6); weak fine and medium subangular blocky structure; friable; slightly sticky; patchy clay films on ped faces; very strongly acid; gradual smooth boundary (9 to 20 inches thick).
B23tg	42-55	Gray (10YR 6/1) sandy clay loam; few medium and coarse distinct mottles of yellowish brown (10YR 5/6); weak fine subangular blocky structure; friable; slightly sticky; patchy clay films on ped faces; very strongly acid; gradual wavy boundary (12 to 18 inches thick).
B3g	55-65+	Gray (10YR 6/1) sandy clay loam; weak coarse subangular blocky structure; friable; patchy clay films on ped faces; very strongly acid.

[8] For a description of symbols used to describe the characteristics of a soil profile, refer to the appendix. The reader should make such reference each time a soil profile description occurs in the remainder of the text.

Table 8   Formative Elements in Names of Suborders

Formative Element	Derivation of Formative Element	Mnemonicon	Connotation of Formative Element
alb	L. *albus*, white.	albino	Presence of albic horizon (a bleached eluvial horizon).
and	Modified from *Ando*.	Ando	Andolike.
aqu	L. *aqua*, water.	aquarium	Characteristics associated with wetness.
ar	L. *arare*, to plow	arable	Mixed horizons.
arg	Modified from argillic horizon; L. *argilla*, white clay.	argillite	Present of argillic horizon (a horizon with illuvial clay).
bor	Gr. *boreas*, northern.	boreal	Cool.
ferr	L. *ferrum*, iron.	ferruginous	Presence of iron.
fibr	L. *fibra*, fiber.	fibrous	Least decomposed stage.
fluv	L. *fluvius*, river.	fluvial	Flood plains.
hem	Gr. *hemi*, half.	hemisphere	Intermediate stage of decomposition.
hum	L. *humus*, earth.	humus	Presence of organic matter.
lept	Gr. *leptos*, thin.	leptometer	Thin horizon.
ochr	Gr. base of *ochros*, pale.	ocher	Presence of ochric epipedon (a light-colored surface).
orth	Gr. *orthos*, true.	orthophonic	The common ones.
plag	Modified from Ger. *plaggen* sod.		Presence of plaggen epipedon.
psamm	Gr. *psammos*, sand.	psammite	Sand textures.
rend	Modified from *Rendzina*.	Rendzina	Rendzinalike.
sapr	Gr. *sapros*, rotten.	saprophyte	Most decomposed stage.
torr	L. *torridus*, hot and dry.	torrid	Usually dry.
trop	Modified from Gr. *tropikos*, of the solstice.	tropical	Continually warm.
ud	L. *udus*, humid.	udometer	Of humid climates.
umbr	L. *umbra*, shade.	umbrella	Present of umbric epipedon (a dark-colored surface.)
ust	L. *ustus*, burnt.	combustion	Of dry climates, usually hot in summer.
xer	Gr. *xeros*, dry.	xerophyte	Annual dry season.

Table 9   Formative Elements for Names of Great Groups

Formative Element	Derivation of Formative Element	Mnemonicon	Connotation of Formative Element
acr	Modified from Gr. *akros*, at the end.	acrolith	Extreme weathering.
agr	L. *ager*, field.	agriculture	An agric horizon.
alb	L. *albus*, white.	albino	An albic horizon.
and	Modified from *Ando*.	Ando	Andolike.
anthr	Gr. *anthropos*, man.	anthropology	An anthropic epipedon.
aqu	L. *aqua*, water.	aquarium	Characteristic associated with wetness.
arg	Modified from argillic horizon: L. *argilla*, white clay.	argillite	An argillic horizon.
calc	L. *calcis*, lime.	calcium	A calcic horizon.
camb	L. *cambiare*, to exchange.	change	A cambic horizon.
chrom	Gr. *chroma*, color.	chroma	High chroma.
cry	Gr. *kryos*, coldness.	Crystal	Cold.
dur	L. *durus*, hard.	durable	A duripan.
dystr	Modified from Gr. *dys*, ill, *dystrophic*, infertile.	dystrophic	Low base saturation.
eutr	Modified from Gr. *eu* good; *eutrophic*, fertile.	eutrophic	High base saturation.
ferr	L. *ferrum*, iron.	ferric	Presence of iron.
frag	Modified from L. *fragilis*, brittle.	fragile	Presence of fragipan.
fragloss	Compound of *fra(g)* and *gloss*.	glossary	See the formative elements *frag* and *gloss*.
gibbs	Modified from *gibbsite*.	gibbsite	Presence of gibbsite.
gloss	Gr. *glossa*, tongue.	glossary	Tongued.
hal	Gr. *hals*, salt.	halophyte	Salty.
hapl	Gr. *haplous*, simple.	haploid	Minimum horizon.

Formative element	Derivation	Mnemonicon	Connotation
hum	L. *humus*, earth.	humus	Presence of humus.
hydr	Gr. *hydro*, water.	hydrophobia	Presence of water.
hyp	Gr. *hypnon*, moss.	hypnum	Presence of hypnum moss.
luo, lu	Gr. *louo*, to wash.	ablution	Illuvial.
moll	L. *mollis*, soft.	mollify	Presence of mollic epipedon.
nadur	Compound of *na(tr)*, and *dur*.		Presence of natric horizon.
natr	Modified from *natrium*, sodium.		Presence of natric horizon.
ochr	Gr. base of *ochros*, pale.	ocher	Presence of ochric epipedon (a light-colored surface.)
pale	Gr. *paleos*, old.	paleosol	Old development.
pell	Gr. *pellos*, dusky.		Low chroma.
plac	Gr. base of *plax*, flat stone.		Presence of a thin pan.
plag	Modified from Ger. *plaggen*, sod.		Presence of plaggen horizon.
plinth	Gr. *plinthos*, brick.		Presence of plinthite.
quartz	Ger. *quarz*, quartz.	quartz	High quartz content.
rend	Modified from *Rendzina*.	Rendzina	Rendzinalike.
rhod	Gr. Base of *rhodon*, rose.	rhododendron	Dark-red colors.
sal	L. base of *sal*, salt.	saline	Presence of salic horizon.
sider	Gr. *sideros*, iron.	siderite	Presence of free iron oxides.
sombr	Fr. *sombre*, dark.	somber	A dark horizon.
sphagno	Gr. *sphagnos*, bog.	sphagnum moss	Presence of sphagnum moss.
torr	L. *torridus*, hot and dry.	torrid	Usually dry.
trop	Modified from Gr. *tropikos*, of the solstice.	tropical	Continually warm.
ud	L. *udus*, humid.	udometer	Of humid climates.
umbr	L. base of *umbra*, shade.	umbrella	Presence of umbric epipedon.
ust	L. base of *ustus*, burnt.	combustion	Dry climate, usually hot in summer.
verm	L. base of *vermes*, worm.	vermiform	Wormy, or mixed by animals.
vitr	L. *vitrum*, glass.	vitreous	Presence of glass.
xer	Gr. *xeros*, dry.	xerophyte	Annual dry season.

Entisols, Vertisols, and Inceptisols

6

The Entisols, Vertisols, and Inceptisols are mineral soils that lack sufficient horizon development and/or chemical alteration to produce the mature soil characteristics associated with their regional climatic environment. In most instances, these soils extend beyond the limits of any particular climatic or floral region. An example is the soils formed in *alluvium* (stream-deposited sediments), like those developed in the valleys of the Mississippi River and its tributaries. Such alluvial soils recur under practically any thermal or moisture regime and under widely varying vegetative covers.

The primary difference among these three soil orders is the degree to which they may approach the status of a mature profile. Entisols exhibit little or no evidence of pedogenic horizons. The Vertisols are dominated by clays with a high shrink-swell capacity and exhibit some mineral alteration. Inceptisols have weakly developed pedogenic features in response to their soil-forming environment.

ENTISOLS

Soil-forming processes in the Entisols either have not been operative for a sufficient period of time or some aspect of the physical environment has prevented the complete development of pedogenic horizons. These are true soils and should not be confused with recently weathered parent material incapable of supporting plant life. Various factors are responsible for the lack of horizon development or mineral alteration. In some, time has been a dominating element. Such Entisols are found on newly exposed surface deposits that have not been in place long enough for pedogenesis to operate to its fullest. Some may be found on steep, actively eroding slopes, others on floodplains, and still others on glacial outwash plains that receive new deposits of alluvium at frequent intervals. There are types, however, that are very old. Consisting primarily of quartz or other minerals that do not alter readily, they simply do not form horizons. Entisols may exist in almost any moisture or temperature region, on any type of parent material regardless of age, and under any form of vegetation (Figure 6.1).

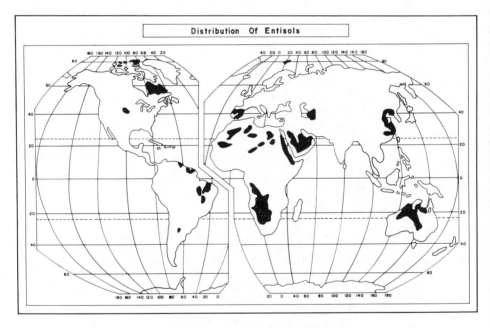

Figure 6.1 World distribution of Entisols.

There are five suborders of Entisols (Figure 6.2):

1. *Aquents* (L. *aqua*, water) are the wet Entisols. They are commonly found in tidal marshes, in deltas on the margins of lakes where the soil is continuously saturated with water, in floodplains of streams where the soil is saturated at some

time of the year, or in very wet sandy deposits. These soils are bluish or gray and mottled. Temperature does not restrict the Aquents, except in areas constantly below freezing. The moisture regime is primarily a reducing one, virtually free of dissolved oxygen due to saturation by ground water or its capillary fringe. Thus, Aquents are generally found in recent, usually water-deposited sediments, but may support any vegetation that will tolerate prolonged periods of wetness.

2. *Arents* (L. *arare*, to plow) are Entisols that lack horizons, normally because of human interference. They have been deeply mixed by plowing, spading, or moving. The Arents retain fragments that can be identified as parts of former diagnostic forms, yet the fragments no longer form continuous horizons. Instead, the remnants are scattered throughout the solum and mixed with other horizons. Some of these soils are the result of deliberate human attempts to either modify the soil or to break up and remove restrictive pans; in other instances they have resulted from cut and fill operations intended to reshape the surface. An uncommon few are formed from the natural effect of mass movement in earth slides. Arents are not extensively developed and their characteristics vary from place to place. Powerful modern machinery seems to be increasing their acreage, and they are likely to become more significant in the future.

3. *Fluvents* (L. *fluvius*, river) are brownish to reddish Entisols. They have formed in recent water-deposited sediments—primarily floodplains, fans, and deltas of rivers and small streams, but not in back swamps where drainage is poor. Although the time span could, in specific instances, be as long as a few hundred years in humid regions, and even longer in arid areas, the age of the sediments in which these soils form is usually very young, a few years or decades. Under normal conditions the Fluvents are flooded frequently and the deposited materials show signs of stratification—layers of a given texture alternate with layers of other textures. Most alluvial sediments, coming from eroding surfaces or stream banks, include appreciable amounts of organic carbon that are dominantly associated with the clay fraction. Hence, clayey or loamy strata usually have more organic carbon than strata that either overlie or underlie them and that are more sandy. The percentages of organic carbon, therefore, decrease irregularly with depth if the

ENTISOLS

Soils that have little or no evidence of development of pedogenic horizons.

AQUENTS

Entisols that exhibit evidence of wetness.

Cryaquents are cold, wet soils of high mountains, tundra, or cold coastal marshes.

Fluvaquents are wet soils of flood plains and deltas of mid and low latitudes.

Haplaquents are wet Aquents in upland depressions where fresh sediments do not accumulate.

Hydraquents are clayey soils of tidal marshes that are permanently saturated with water.

Psammaquents have sandy textures and gray or mottled gray colors.

Sulfaquents have sulfidic materials within 20" of the surface.

Tropaquents are permanently warm and wet soils in depressions of intertropical regions.

ARENTS

Entisols that lack horizons because they have been deeply mixed by plowing, spading, or moving by man. (There are no great groups.)

FLUVENTS

Entisols formed in recent water-deposited sediments, primarily in floodplains, fans, and deltas of rivers and small streams, but not in back swamps where drainage is poor.

Cryofluvents have a cryic temperature regime.

Torrifluvents are found in arid climates and are not flooded frequently or for long periods.

Tropofluvents have a udic moisture regime and an isomesic or warmer iso-temperature regime.

Udifluvents have a udic moisture regime and frigid to hyperthermic temperature regime.

Ustifluvents have a ustic moisture regime and mesic or isomesic or warmer temperature regimes.

Xerofluvents have a xeric moisture regime.

ORTHENTS

Entisols on recent erosional surfaces.

Cryorthents are found in high mountains or high latitudes.

Torriorthents are the dry and salty Orthents of cool to hot, arid regions.

Troporthents are found in intertropical regions with udic moisture regimes.

Udorthents have a udic moisture regime and are found in the midlatitudes.

Ustorthents have a ustic moisture regime and are found in mid- or low latitudes.

Xerorthents have a xeric moisture regime.

PSAMMENTS

Entisols in poorly graded sands of shifting or stabilized sand dunes, cover sands, or in parent materials sorted in an earlier geologic cycle.

Cryopsamments are found in cold places.

Quartzipsamments are freely drained quartz sands of humid to semiarid regions.

Torripsamments are found in arid climates.

Tropopsamments have some weatherable minerals, a udic moisture regime, and an iso-temperature regime.

Udipsamments are found in humid climates within the midlatitudes.

Ustipsamments have a ustic moisture regime.

Xeropsamments are found in Mediterranean climates.

Figure 6.2 Suborders and Great Groups of the soil order Entisol.

materials are stratified. If textures are homogeneous, on the other hand, the organic carbon content will decrease regularly with depth. These soils do not occur under any specific vegetation and may be found in any moisture or thermal regime, except those that are subject to temperatures constantly below freezing.

4. *Orthents* (Gr. *orthos*, true) are Entisols occurring primarily on recently eroded surfaces that have been created by geologic factors or produced by cultivation. The basic requirement is that any former existing soil has been either completely removed or so truncated that the diagnostic horizons typical of all orders other than Entisols are absent. In some cases remmants of indurated diagnostic horizons, such as ironstone that once may have been plinthite,[1] may be present if exposed at the surface. Such formations normally support only scattered plants. If they do not support plants, they are considered to be rock rather than soil. A few Orthents occur in recent loamy or fine wind-deposited sediments, glacial deposits, debris from recent landslides and mudflows, and in recent sandy alluvium.

5. *Psamments* (Gr. *psammos*, sand) are sandy Entisols that lack pedogenic horizons. They include sandy dunes, cover sands, and sandy parent material produced in an earlier geologic cycle. Others are found in sands that have been sorted by water, natural levees or beaches. Occurring under any climate or vegetation, they may be located on surfaces of virtually any age. The older Psamments are usually dominated by quartz sand and cannot form subsurface diagnostic horizons that involve the accumulation of clays and sesquioxides.

A profile description of a Psamment is shown below. This soil is found in Calhoun County, Florida, and is a member of the thermic, coated family of Typic Quartzipsamments. These soils have thin, very dark grayish brown, sandy A horizons and yellowish brown, sandy C horizons.

VERTISOLS

Vertisols are clayey soils that have deep, wide cracks during periods of moisture deficiency. The clays making up these soils swell

---

[1] Plinthite is discussed in detail in Chapter 11.

Horizon	Depth (Inches)	Typifying Pedon: Lakeland Sand
A	0-3	Very dark grayish brown (10YR 3/2 crushed) sand; single grained; loose; moisture of very dark grayish brown (10YR 3/2) organic matter and clean, uncoated, white (10YR 8/1) sand grains; common fine and medium roots; strongly acid; clear wavy boundary (2″ to 8″ thick).
C1	3-10	Yellowish brown (10YR 5/4) sand; common splotches of yellowish brown (10YR 5/6); single grained; loose; common fine and medium roots; few uncoated sand grains; strongly acid; gradual wavy boundary (4″ to 10″ thick).
C2	10-43	Yellowish brown (10YR 5/8) sand; single grained; loose; few fine roots; few uncoated sand grains; strongly acid; gradual wavy boundary (30″ to 50″ thick).
C3	43-64	Yellowish brown (10YR 5/8) sand; few medium faint splotches of very pale brown (10YR 7/3, 7/4); single grained; loose; many uncoated sand grains; strongly acid; gradual wavy boundary (18″ to 30″ thick).
C4	64-90	Very pale brown (10YR 7/4) sand; few medium distinct yellowish red (5YR 4/8) mottles; single grained; loose; many uncoated sand grains; strongly acid.

upon wetting and shrink when dried. In the United States the dominant clay mineralogy is montmorillinite, although in other parts of the world illite or a mixed mineralogy may produce the same characteristics.

The formation of Vertisols involves the production of 2:1 layer-lattice, expanding clays by weathering activity within a climatic realm that experiences strong periodic contrasts in its moisture supply (Figure 6.3).[2] The source of the clay may be argillaceous sedimentary materials or the result of the chemical alteration of basic igneous products. In either case, maintaining the soil's shrink-swell properties depends upon whether or not further weathering destroys the clay minerals' expanding character.

Upon partial drying the clays shrink and produce wide deep cracks. These open cracks may be in excess of 1″ wide to a depth of 20″ (50cm). In certain areas cracks have been known to extend to depths in excess of 40″ (1 meter). By definition, an open crack is considered a separation between very coarse prisms or polyhedrons. The open cracks permit surface materials that are displaced by precipitation, animals, or the like to fall to lower depths of the solum within the confines of the separations. When rains occur, water runs

[2] In arid regions they commonly develop in closed depressions or playas that are occasionally flooded or in fine-textured materials in areas that have infrequent heavy showers. In other regions seasonal moisture changes are chiefly the result of excesses of evapotranspiration over precipitation in one season, followed by an excess of precipitation in the next.

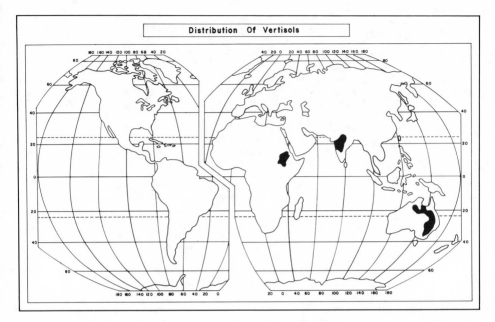

Figure 6.3 World distribution of Vertisols.

into the cracks readily; hence the soil is remoistened both from above and below. With moistening, the clays expand and the cracks close, trapping the displaced particles at lower levels (Figure 6.4). As the profile moistens from below, the clays in lower horizons tend to swell before those above them. This results in one part of the soil moving against another as pressure increases due to an increased volume of material in the lower portion of the solum. Pressure is exerted in all directions, but the soil is only capable of moving horizontally or upward. It is thought that this process is the probable means of producing *slickensides* and *gilgai.* Slickensides are polished and grooved surfaces in the soil made by one mass of material sliding past another. Gilgai is a microrelief feature consisting of microbasins and microknolls, and represents a surface warping in response to internal pressure differentials.

The Soil Survey staff recognizes four suborders of Vertisols (Figure 6.5):

1. *Torrerts* (L. *torridus*, hot, dry) are the Vertisols of arid climatic regions. They have cracks that can remain open the entire year, although they may be partially or largely filled with a mulch that has been moved by either wind or animals. If the area has a short rainy period and the cracks close, they remain closed for less then 60 consecutive days.

VERTISOLS

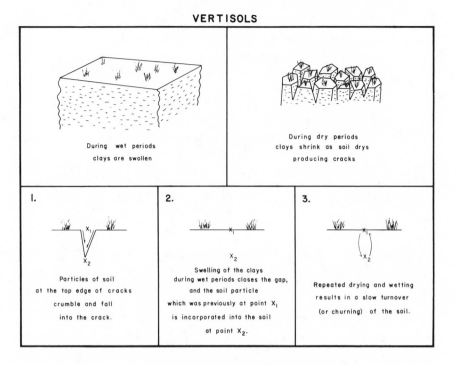

Figure 6.4   Vertisol development.

2. *Uderts* (L. *udus*, humid) are the Vertisols of humid climatic realms. Their cracks open and close at irregular times, depending upon the weather. In fact, some years may experience no cracking at all, although this must occur in less than 50 percent of all years.

3. *Usterts* (L. *ustus*, burnt) are the Vertisols of tropical and subtropical monsoon climates that experience two distinct rainy and dry seasons, and of temperate regimes that are characterized by limited summer rains. These soils have cracks that open once or twice during the year and remain open for 90 cumulative days or more in most years. The cracks are normally closed for 60 consecutive days or more each year during the wet season.

4. *Xererts* (Gr. *xeros*, dry) are the Vertisols of the earth's Mediterranean climates, for example, areas with cool, wet winters and warm to hot, arid summers. They have cracks that close and open regularly once per year in association with the alternating wet and dry seasons. In most summers the cracks will remain open for more than 60 days.

VERTISOLS

These are clayey soils that have deep wide cracks at some time during the year.

TORRERTS

Vertisols of arid climates. They have cracks that may remain open throughout the year or are closed for less than 60 days.

UDERTS

Vertisols of humid climates. They have cracks that open and close 'irregularly over time, according to the weather.

Chromuderts have a readily visible color other than black, gray, or white.

Pelluderts are dominantly gray to black in all subhorizons.

USTERTS

Vertisols of monsoon climates, of tropical and subtropical areas that have two rainy and two dry seasons, and of temperate regions that have limited summer rains.

Chromusterts have a readily visible color other than black, gray, or white.

Pellusterts are dominantly gray to black in all subhorizons.

XERERTS

Vertisols of Mediterranean climates, with cool wet winters and warm dry summers.

Chromoxererts have a readily visible color other than black, gray, or white.

Pelloxererts are dominantly gray to black in all subhorizons.

Figure 6.5 Suborders and Great Groups of the soil order Vertisol.

INCEPTISOLS

The Inceptisols include soils of widely differing environments in which a variety of pedogenic processes are operative. Some are weathered sufficiently to produce altered horizons that have lost either bases or iron and aluminum, but still retain weatherable minerals. At the same time they normally lack illuvial horizons that have been enriched with either silicate clays containing aluminum or amorphous mixtures of aluminum and organic carbon. Smith refers to these soils as "primarily eluvial," that is, losing material throughout the profile.[3] Inceptisols are generally found in humid climates where leaching is active (Figure 6.6). In short, Inceptisols are beginning to exhibit pedogenic characteristics associated with weathering, but these features are too weak to be considered maturely developed.

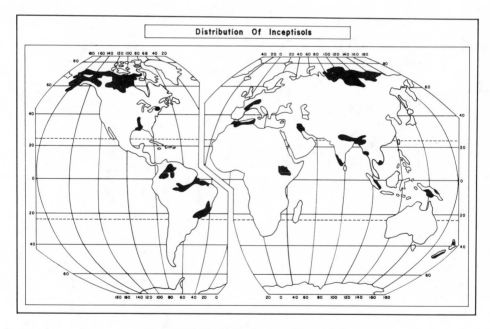

Figure 6.6 World distribution of Inceptisols.

There are six suborders of Inceptisols (Figure 6.7):

1. *Andepts* are relatively freely drained. They have a low bulk density and are usually formed in pyroclastic materials (such

[3] Guy D. Smith, "Lectures on Soil Classification," *Pedologie,* IV (Ghent, Belgium: State University of Ghent, 1965), 33.

INCEPTISOLS

These soils have altered horizons that have lost bases or iron and aluminum but retain weatherable minerals and that lack illuvial horizons either enriched with silicate clays that contain aluminum, or those enriched with amorphous mixtures of aluminum and organic carbon.

ANDEPTS

Inceptisols that are relatively freely drained and have appreciable amounts of allophane or pyroclastic materials.

Cryandepts are the cold Andepts of high mountains or high latitudes.

Durandepts have a duripan within 40″ of the surface.

Dystrandepts have large amounts of organic carbon and amorphous materials, and a small amount of bases.

Eutrandepts have large amounts of organic carbon and amorphous materials, and an ample supply of bases.

Hydrandepts have perudic moisture regimes.

Placandepts have placic horizon.

Vitrandepts have the largest amounts of vitric ash and pumice and the lowest amounts of their weathering products.

AQUEPTS

These are wet Inceptisols with poor drainage.

Andaquepts are formed in pyroclastic materials.

Cryaquepts are found in arctic and subarctic regions.

Fragiaquepts have a fragipan.

Halaquepts are sodic and saline soils.

Haplaquepts are light colored, gray Aquepts, mostly of humid climates.

Humaquepts are nearly black or peaty, very wet, acid Aquepts.

Placaquepts have a placic horizon.

Plinthaquepts have plinthite present.

Sulfaquepts are acid sulfate soils.

Tropaquepts are found in the tropics.

OCHREPTS

Inceptisols that are freely drained and light in color.

Cryochrepts are found in the cold areas of high mountains or high latitudes.

Durochrepts have a duripan in the upper 40″.

Dystrochrepts are brownish, acid soils of humid and perhumid regions.

Eutrochrepts are brownish, base rich soils of humid regions.

Fragiochrepts have a fragipan.

Ustochrepts are found in subhumid to semiarid regions.

Xerochrepts are found in Mediterranean climates.

PLAGGEPTS

Inceptisols that have a plaggen epipedon composed of crystalline rather than pyroclastic materials.

TROPEPTS

Inceptisols that are freely drained, brownish to reddish, and found in intertropical regions.

Dystropepts are acid Tropepts of high rainfall regimes.

Eutropepts have high base saturation and are rarely dry for extended periods.

Humitropepts are found in cool humid regions of high altitude.

Sombritropepts are dark, humus rich Tropepts of perhumid cool, hilly or mountainous regions.

Ustropepts are base rich Tropepts of subhumid regions.

UMBREPTS

Inceptisols which are acid, dark reddish or brownish, freely drained, and organic rich.

Cryumbrepts are found in high latitudes or high altitudes.

Fragiumbrepts have a fragipan.

Haplumbrepts are freely drained and have either a short or no dry season during the summer.

Xerumbrepts are found in Mediterranean climatic regimes.

Figure 6.7 Suborders and Great Groups of the soil order Inceptisol.

as volcanic ash or pumice), although certain sedimentary and basic extrusive igneous rocks can also serve as the parent material. Andepts are rich in either glass or allophanes, such as amorphous clays. Glass is a unique parent material in that it is relatively soluble in comparison to the crystalline aluminosilicates, and weathers rapidly to produce amorphous products. Found in any latitude, the location of these soils is restricted to areas in or near mountains with active volcanoes. Repeated ash falls are quite common in Andepts, and these soils may have two or more buried horizons within 40″ (1 meter) of the surface.

2. *Aquepts* (L. *aqua*, water) are wet Inceptisols. The natural drainage of these soils is poor or very poor and the ground water table usually stands close to the surface at some time during the year. The surface horizons are dominantly gray to black, and subsurface horizons are dominantly gray.

3. *Ochrepts* (Gr. *ochros*, pale) are light-colored, brownish, and relatively freely drained Inceptisols of middle to high latitudes. They are formed in crystalline parent materials and in a moisture regime with an annual excess of precipitation over evapotranspiration.

4. *Plaggepts* (Ger. *plaggen*, sod) include all freely drained soils that contain a plaggen epipedon, except for a few Andepts:

The plaggen epipedon is a man-made surface layer, more than 50cm thick, that has been produced by long-continued manuring. In medieval times, sod or other materials were commonly used for bedding livestock and the manure was spread on the field being cultivated. The mineral materials brought in by this kind of manuring eventually produced an appreciably thickened Ap, as much as 1 meter or more in thickness.

Colors and contents of organic carbon depend on the sources of the materials used for bedding. If the sod was cut from the heath, the plaggen epipedon tends to be black or very dark gray, to be rich in organic matter, and to have a wide carbon-nitrogen ratio. If the sod came from forested soils, the plaggen epipedon tends to be brown, to be lower in organic matter, and to have a narrower carbon-nitrogen ratio.

The plaggen epipedon may be identified by several means. Commonly it contains artifacts, such as bits of brick and pottery throughout. Chunks of diverse materials, such as black and light gray sand as large as the size held by a spade, may be present. The plaggen epipedon normally shows spade marks throughout as well as remnants of thin stratified beds of sand that were probably produced on the surface by beating rains and later buried by spading. The polypedons that have plaggen epipedons tend to be straight-sided rectangular bodies, and they are usually higher than adjacent polypedons by as much or more than the thickness of the plaggen epipedon.[4]

5. *Tropepts* (Gr. *tropikos*, of the solstice) are found within the tropics and include freely drained Inceptisols that are brownish to reddish in color. They normally have an ochric epipedon or a cambic horizon and lack significant amounts of active amorphous clays or pyroclastic materials. Within this soil's climatic regime:

Biologic activity is continuous unless there is a pronounced dry season. Cultivated soils cannot remain bare and exposed to erosion for long periods, for if there is rain the plants start to grow. The continuous biologic activity is reflected in the nature of the organic matter in at least two important ways. One is that for any given kind of soil, C-N ratios tend to be lower in tropical than temperate regions. C-N ratios of virgin soils in the humid tropics compare with those of Aridisols in temperate regions; often they are extremely low. The other difference is that the amount of organic matter is not well reflected by the color of the soil. Black soils may have little organic matter; light-colored soils may be very rich in organic matter, or very poor.[5]

6. *Umbrepts* (L. *umbric*, shade) are acidic, dark reddish or brown, freely drained, organic rich Inceptisols that occur in humid regions of the middle to high latitudes. These soils have much more organic matter than the Ochrepts, but their base saturation is too low to classify them as Mollisols. They occur in hilly to mountainous regions with relatively high precipitation, even though many experience a distinct dry season during the summer.

LAND-USE AND MANAGEMENT PROBLEMS

Because they occur in varied sites and climates on a multitude of parent materials, Entisols, Vertisols, and Inceptisols have many uses and pose many diverse management problems. The most significant of these soils, in terms of areal occurrence, are the

[4] National Cooperative Soil Survey, *Soil Taxonomy*, pp. 3-6.
[5] Smith, "Soil Classification," p. 41.

Vertisols, the Entisol suborders Aquents, Fluvents, and Psamments, and the Inceptisol suborder Aquepts. Just as water is the primary agent in the formation of the aquic and fluvial suborders (Aquents, Aquepts, and Fluvents), its management is also a major factor in making these soils agriculturally productive. Under natural conditions most have supported a forest cover of water-tolerant trees. When the lands are cleared for agricultural purposes, flooding and poor drainage pose a constant threat to the stability of the crop yields and are the chief causes of crop failure unless surface and subsurface drainage systems and flood control measures are effectively employed to remove and regulate the surplus water.

Crops grown on these immature soils include: cotton, soybeans, corn, oats, wheat, barley, sugarcane, rice, and vegetables. The success of crop production after drainage is largely dependent upon the physical condition of the soil and the availability of nutrients for plants.

The primary factors that are detrimental to the soil's physical properties are: loss of organic matter, erosion, inadequate drainage, improper management practices, and equipment traffic.[6] In areas of relatively high temperatures and surplus moisture (such as the Mississippi Delta region), organic materials rapidly decompose when aerated. As a consequence, such soils lose a major portion of their original organic matter shortly after being brought under cultivation. This results in reduced stability of soil structure, decreased water infiltration, and increased surface runoff. The same conditions responsible for the depletion of organic material also create difficulties in replenishing or maintaining this soil component.

Although the landscapes of the aquic and fluvial environments normally have gentle slopes, erosion can be a serious problem.

> Most of the organic matter was in the top few inches and has been most affected by erosion. Furthermore, the ratio of sand, silt, and clay in the surface soil provided better physical conditions than the deeper soil. The removal of the surface layer by sheet erosion leaves material with less desirable physical properties exposed or near the surface.[7]

Man's manipulation of these soils may also lead to their deterioration. If cultivated when too wet, soil structure may break down. In other instances, *pressure pans* may develop from the

---

[6] The following summary of land-use and management problems is largely derived from Perrin H. Grissom, "The Mississippi: Delta Region," *The 1957 Yearbook of Agriculture: Soil* (Washington, D.C.: U.S. Govt. Ptg. Office, 1957), pp. 524-30.

[7] Ibid., p. 527.

movement of heavy equipment over wet ground.[8] The resulting compacted layer resists the vertical movement of water and root penetration. Consequently, plants may suffer from too much water during rainy periods and from a lack of moisture during interprecipitation periods. One possible solution is *deep tillage*, or subsoiling, which temporarily eliminates the compacted zone. When deep tillage is practiced on compacted soils, the water infiltration rate, field capacity, and root development all increase and crop yields are generally higher.

Many farmers, in an attempt to improve the physical condition of their soil, have grown winter legumes, which they plow under prior to planting a summer crop. This reduces erosion, and increases organic matter and available nitrogen.

As is true with all soils, the availability of nutrients to meet plant demands is an additional factor in the management of the aquic and fluvial lands. The usual primary deficiencies are nitrogen, phosphorus, potassium, and lime. Nitrogen is required for practically all nonleguminous crops, and may be supplied either by crop rotations that include legumes or by commercial inorganic materials. Generally, it is more economical to use commercial nitrogen on cultivated crops.

Psamments may be stabilized sand dunes or soils developed on sandy deposits. In most instances they are of extremely limited agricultural use. Due to the soil's low moisture-holding capacity and low organic content, coupled with a deficient nutrient supply, potential crop yields are usually low. Psamments' primary use has been for livestock grazing, although in favored climates some soils have been modified by man and now support truck farms and citrus groves.

Vertisols have their maximum occurrence between latitudes 45° North and South, and are most extensively developed in Australia, India, and the Sudan.

Agriculturally the soils have great potential where power tools, fertilizers, and irrigation are available. The natural fertility level can be considered quite high, although the use of nitrogen and phosphorus is beneficial. Tillage of the soil is difficult with primitive tillage tools. The "blacklands" of Texas and Alabama are some of the best agricultural lands in the

---

[8] A pressure pan is a subsurface horizon or soil layer having a higher bulk density and a lower total porosity than the soil directly above or below it, as a result of pressure that has been applied by normal tillage operations or by other artificial means. It is frequently referred to as plow pan, plow sole, and traffic pan.

United States. Worldwide Vertisols are used mainly for cotton, wheat, corn, sorghum, rice, sugarcane, and pasture.[9]

The major problems with using Vertisols relates to their high shrink-swell capacity. The stress created by the alternate expansion and contraction of their clays has been known to crack building foundations and road surfaces, and to be a menace to grazing animals that may be injured by stepping into cracks in the soil.

---

[9] H. D. Foth and L. M. Turk, *Fundamentals of Soil Science*, 5th ed. (New York: John Wiley and Sons, Inc., 1972), p. 266.

Aridisols

7

Aridisols are mineral soils with an ochric (pale) epipedon. These soils may contain one of a wide variety of diagnostic subsurface features. Table 10 defines the most common of these pedogenically related forms. The spatial occurrence of these soils is closely related to the climatic deserts of the world (Figure 7.1). Notably, large expanses are found in the Sahara, Namib, Atacama, Sonoran, Australian, Thar, and Gobi deserts. These soils have the shallowest profiles of those regionally developed at the order level. Their depth characteristics and pale color are significantly associated with a climate of high moisture demand and low water supply and sparse vegetative cover.

CLIMATE AND VEGETATION

Arid climatic regimes occupy approximately 20 percent of the world's land area. These regions have a large disparity between potential evapotranspiration (PE) and environmental water supply. A high incidence of clear skies permits solar radiation to reach land surfaces with a minimum of depletion. Thus, the desert surface heats very rapidly after sunrise, raising ground temperatures to high levels and creating very steep, near-surface, atmospheric lapse rates. After sunset, the reverse is true; the clear desert skies permit a rapid loss of energy through terrestrial radiation, and temperatures drop quickly. As a consequence, temperature patterns exhibit high ranges and

Table 10   Significant Features of Soil Classification in the Aridisol Order

Epipedon:

Ochric	surface horizons that are light in color and low in organic material.

Diagnostic Horizons:

Argillic	a subsurface illuvial horizon of the mineral soil in which layer-lattice silicate clays have accumulated by illuviation to a significant extent.
Calcic	horizons of secondary carbonate enrichment that are $>6''$ thick and have a calcium carbonate equivalent content >15 percent, and have $\geqslant 5$ percent calcium carbonate than the underlying C horizon.
Cambic	a mineral soil horizon that has a texture of loamy very fine sand or finer, has soil structure rather than rock structure, contains some weatherable minerals, and is characterized by the alteration or removal of mineral material as indicated by mottling or gray colors, stronger chromas or redder hues than in underlying horizons, or the removal of carbonates.
Gypsic	a subsurface horizon that is enriched with calcium sulfate, containing at least 5 percent more gypsum than the C horizon.
Natric	a special kind of argillic horizon that has the properties of: (1) prismatic or columnar structure, or occasionally blocky structure; and (2) in some subhorizon has at least 15 percent saturation with exchangeable sodium.
Petrocalcic	a continuous indurated calcic horizon, cemented by carbonates of calcium, and in places with some magnesium carbonate.
Salic	a subsurface horizon at least $6''$ thick, enriched in soluble salts, generally at least 2 to 3 percent depending on thickness.

Other Diagnostic Features:

Duripan	a subsurface horizon that is cemented by silica, usually opal or microcrystalline forms, to the point that fragments from the air-dry horizon will not slake in water or acid.
gilgai	microrelief features of soil produced by expansion and contraction with changes in moisture. Found in soils containing large amounts of clay that shrink and swell considerably with wetting and drying. Usually has the appearance of a succession of microbasins and microknolls.
lithic contact	a boundary between soil and continuous coherent underlying material with a hardness > 3 Mohs.
paralithic contact	similar to lithic contact except that the underlying material is softer (< 3 Mohs).
slickensides	shiny ped surfaces created in clay soils by one mass of soil sliding past another.

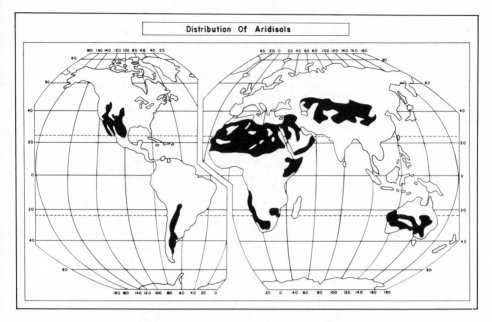

Figure 7.1   World distribution of Aridisols.

maxima, both diurnally and seasonally.[1] The amplitude of the
temperature curve in Figure 7.2 is based on mean monthly data and
is typical of an arid, continental location. The related moisture
demand (PE) is also shown in the same diagram. The amplitude of
the moisture demand curve is proportionately much higher than that
for temperature. This indicates that temperature increase basically
behaves in a linear fashion, while the capacity of the atmosphere to
evapotranspire and store moisture increases at an increasing rate, that
is, exponentially (Figure 7.3). Thus, it can be seen that with
temperature maxima usually above the norm for their latitudinal
location, the atmosphere of desert regimes possesses high energy
availability. The relationship between this energy available for
evapotranspiring moisture and the environmental water supply is a
critical factor in determining an area's degree of aridity.

Moisture supply in arid regions is erratic, and mean precipitation figures provide but scant information on expected amounts of
available water during any given period. In a 10-year span at Las
Vegas (Table 11), for example, annual precipitation ranged from .56"
to 5.55". This is a variation from the annual precipitation norm

---

[1] The primary exception is found in coastal deserts where cool ocean
currents may modify temperature extremes considerably.

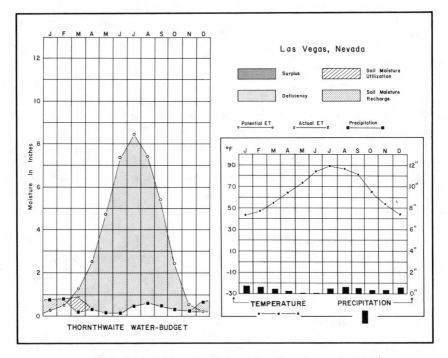

Figure 7.2 Water Budget: Las Vegas, Nevada (based on normal data).

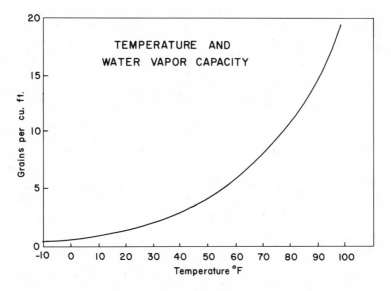

Figure 7.3 Relationship between temperature and water vapor holding capacity of the atmosphere.

(3.90″) of 14 and 142 percent respectively. Individual months experience even greater departures. August has a percentage range from less than 2 to over 539 percent.

The moisture that does reach the surface is not only unreliable and low in total amount, but also generally has minimal opportunity to infiltrate the soil and percolate through the profile. Desert landscapes that have been baked by the sun during interprecipitation periods are not often receptive to water infiltration. Furthermore, much of the precipitation in these areas arrives as thermally induced convective showers of short duration. Hence, surface water either: (1) becomes rapidly channelized and infiltrates into the more porous stream beds to recharge ground-water storage, or (2) is lost through evaporation on the hot desert surfaces under conditions of high energy availability. This brief, transient role of surface water in the desert has led numerous researchers to believe that the profile development and mineral alteration of many Aridisols may be partially the product of past climates that were characterized by higher, more uniform rainfall. A re-examination of Figure 7.2 shows that there is little soil moisture available for leaching processes, which lends support to such thinking.

Stressing the uniqueness of the desert regime, it must be understood that aridity is strictly a phenomenon of climate, not of vegetation or soil formation—it represents deficient moisture and must include the limits of water demand as well as supply. Too frequently, simple measurements of aridity rely solely on an area's precipitation record, which results in an erroneous delimitation of climatic deserts. Such an example would be the inclusion of the Antarctic Plateau, where precipitation is about 4″ to 5″, in a desert classification. Obviously, such a designation would be ridiculous, even though the area does, in fact, receive very little moisture. One need only question how glacial ice, which exists exclusively in areas of surplus moisture, could accumulate to depths of thousands of feet if the plateau was, indeed, a moisture-deficient, or arid, region. The obvious answer is that glacial ice could not form unless moisture demands were low and a surplus existed. Thus, while such regions may be deserts in the sense that large portions of them cannot support plant life, they are definitely not climatic deserts.

It is also incorrect to conceive of all deserts as being composed of active, shifting sand dunes. Such sandy ergs[2] make up a minor portion of the world's deserts, less than 25 percent. Practically all of

---

[2] Unstablized sand dunes do not exhibit profile development and are not classified in the Aridisol order; rather, they are identified as Entisols.

Table 11 Precipitation Data for the Period 1951-60 and Precipitation Normal Values for Las Vegas, Nevada.

Year	J	F	M	A	M	J	J	A	S	O	N	D	Annual
1951	.19	.02	.03	.04	.10	.00	.54	.16	.98	.05	.44	.23	2.78
1952	1.07	T	1.50	.57	T	.01	.47	.49	.87	.00	.16	.41	5.55
1953	.01	.00	.03	T	.03	T	.23	.10	.02	.12	.02	T	.56
1954	.91	.02	.81	T	.00	.01	1.61	.28	.45	.02	.36	.24	4.71
1955	1.40	.13	T	.11	.03	.39	1.55	1.74	.00	T	.03	.02	5.40
1956	.23	.00	.00	.08	T	.00	1.64	.00	.00	.09	.00	.00	2.04
1957	.26	.16	.10	.55	.15	T	.41	2.59	T	.49	.26	.01	4.98
1958	.43	.73	.55	.64	.18	.00	.01	.18	.21	.63	.96	.00	4.52
1959	.11	.72	T	T	T	T	.02	.33	.01	.51	1.09	1.38	4.17
1960	.56	.42	.05	.10	T	T	.41	T	.23	.49	1.88	.26	4.40
Normal	.53	.44	.35	.23	.08	.04	.50	.48	.34	.20	.31	.40	3.90

the remaining 75 percent of desert land contains some vegetation. Those plants that do exist in desert environments have a wide variety of techniques for surviving in moisture-deficient areas. One characteristic that is common to all arid land vegetation, however, is that the plants are widely spaced. In general, the greater the intensity of dryness, the more widely separated are the individual plants. Thus, there is normally a decrease in total biomass (amount of living matter in a given area) associated with decreased moisture. This situation has important effects on the soil. Its color, structure, microbial population, exchange capacity, and nitrogen content are all heavily influenced by the amount of organic matter that is present.

PEDOGENESIS

Of all the regionally distributed soils, Aridisols exhibit the least depth of horizon and total solum development. Their formation in an arid climate has meant that certain types of chemical weathering are limited and excessive leaching of soluble minerals, which would otherwise be removed from the profile, is prevented.

Two groups of soil-forming processes predominate in the arid regime. One is *calcification*, which is dominant over extensive segments of the earth's moisture-deficient regions. The other is *salinization*, which is normally localized in area.

Calcification

The processes by which calcium carbonate or calcium and magnesium carbonates may accumulate in the soil profile are called calcification. These processes result in the production of either a *calcic horizon* or a *petrocalcic horizon*, the latter known as *caliche* in the southwestern United States. This feature, designated as the *ca* layer in profile description, is normally found in the C horizon, but it may encroach into the uppermost horizon under intensely arid conditions.

A calcic horizon is a subsurface soil layer of secondary carbonate enrichment at least 6″ thick. The primary genetic factor seems to be limited rainfall that is insufficient to remove lime completely from even the few surface inches of the soil. If the parent material is rich in carbonates or is "wind-dusted" with regular carbonate additions, the calcic horizon through time will become plugged with carbonates and will be cemented into a hard, massive, continuous petrocalcic horizon. The petrocalcic horizon is normally restricted to older landscapes; the young calcic horizons have lime accumulations that are soft and disseminated, or else concentrated in small, hard concretions.

A model representing the current interpretations of caliche formation is shown in Figure 7.4. Precipitation falling through the atmosphere dissolves free carbon dioxide, producing a mild carbonic acid. This solution infiltrates the unsaturated portion of the soil, leaching the surface materials and taking ions of the very soluble calcium into solution. The depth to which this soil solution can percolate depends on several variables, including the amount, intensity, and frequency of rainfall; the permeability of the soil complex; and the moisture demands of the environment. Jenny and Leonard have demonstrated that increasing precipitation totals also increase the depth at which carbonate accumulation takes place.[3]

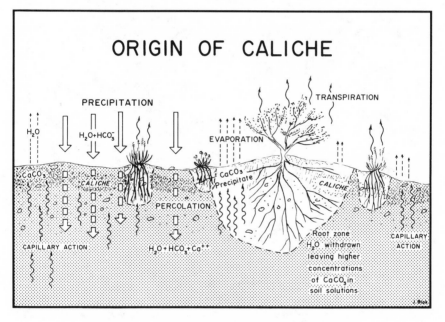

Figure 7.4   A schematic model of Caliche development.

Accompanying each succeeding precipitation period, the calcium carbonate in the accumulation zone increases. Interprecipitation periods, on the other hand, show a loss of moisture from the solum through evapotranspiration. Consequently, the soil solution becomes concentrated with soluble minerals to the saturation point. Further

[3] Hans Jenny and C. D. Leonard, "Functional Relationships Between Soil Properties and Rainfall," *Soil Science*, 38, 363-81.

moisture withdrawals from this saturated solution result in a precipitation of minerals, most frequently calcium carbonates. Subsequent rains continue the processes and result in a thickening of the carbonate zone in the direction of the land surface. Such accretion will end only when the supply of calcium ions is no longer sufficient to develop the caliche layer further. When this stage of weathering is achieved, the uppermost surface of the caliche is repeatedly dissolved and reprecipitated, resulting in a hard crust on the surface of the caliche horizon.

Occasionally very thick deposits of caliche are recorded. These are considered to be the product of soils that are experiencing constant alluviation, hence having available a continuing supply of fresh minerals for the weathering and release of ions. A profile description of an Aridisol with a petrocalcic horizon follows:

Horizon	Depth (Inches)	Cave Gravelly Sandy Loam[4] (Pima County, Arizona)
A1	0-1	—brown (7.5 YR 5/4) gravelly sandy loam, dark brown (7.5 YR 4/4) moist; weak thick platy structure; slightly hard, very friable, nonsticky, nonplastic, common very fine and fine roots; common fine and very fine interstitial and vesicular and few fine tubular pores; 20 percent gravel; strongly effervescent; moderately alkaline (pH 8.0); clear smooth boundary.
C1	1-7	—light brown (7.5 YR 6/4) gravelly fine sandy loam, dark brown (7.5 YR 4/4) moist; massive; slightly hard, very friable, slightly sticky, slightly plastic; common very fine, fine, and medium roots; few fine vesicular and interstitial pores; 30 to 35 percent gravel; strongly effervescent; moderately alkaline (pH 8.2); abrupt wavy boundary.
C2cam	7-20	—white (10 YR 8/2) gravelly indurated layer with thin laminar layer on upper surface, very pale brown (10 YR 7/3) and light yellowish brown (10 YR 6/4) moist; massive; extremely hard; very few very fine roots; few fine vesicular pores; violently effervescent; moderately alkaline (pH 8.2); abrupt wavy boundary.
C3cam	20-32	—white (10 YR 8/2) and very pale brown (10 YR 8/3) strongly cemented gravelly layer, very pale brown (10 YR 7/3) and light yellowish brown (10 YR 6/4) moist; massive; extremely hard to very hard, very firm; very few very fine roots; violently effervescent; moderately alkaline (pH 8.2); clear wavy boundary.
C4cam	32-41	—white (10 YR 8/2) and very pale brown (10 YR 8/3) strongly cemented gravelly layer, very pale brown (10 YR 7/3) and light yellowish brown 10 YR 6/4) moist; massive; very hard to extremely hard; very firm; very few very fine roots; violently effervescent; moderately alkaline (pH 8.2).

[4] D. M. Hendricks and Y. H. Havens, *Desert Soils Tour Guide*, mimeographed pamphlet prepared for the Soil Science Society of America (August 27, 1970), pp. 24-25.

Salinization

The processes of salinization lead to the accumulation of mineral salts in the soil profile of sufficient concentration to limit plant production to those species with a high salt tolerance, for example halophytes. In arid regions, salinization is closely associated with extremely restricted surface water removal in localized areas of entrapped drainage. It is common for such a condition to exist in *bolsons*,[5] where precipitation can become impounded in valley bottoms. These enclosed basins, with no exiting stream channels, would become permanent lakes in humid regions. In arid climates, water descending from surrounding slopes accumulates soluble minerals and transports them into the basin, where an intermittent lake may be formed during the precipitation period. During interprecipitation periods, the entrapped water evaporates, leaving a residue of mineral salts encrusted on the surface and precipitated within the soil profile. These mineral salts can produce salic or natric horizons. A *salic horizon* must be at least 6″ thick and have a minimum of 2 percent soluble salts. The *natric horizon* is a special type of *argillic* (clay) *horizon*. In addition to the properties of the argillic horizon, it must also have: (1) prismatic or columnar structure, and (2) in some portion of the horizon a minimum of 15 percent saturation with sodium. Thorne and Seatz describe the accumulation of mineral salts and their effect on the soil as follows:

> The soluble salts that accumulate in soils consist principally of cations of calcium, magnesium, and sodium and the chloride and sulfate anions. Potassium, bicarbonates, and nitrate ions occur in smaller quantities. Borates occasionally occur in small amounts but receive considerable attention because of their exceptionally high toxicity to plants. The proportion of ions occurring in soils varies greatly. The nature of the salts present obviously depends on the composition of the rocks weathered, the nature of the weathering process, and subsequent reactions during the moving of the salts from the site of weathering to the place of deposition.
>
> Salt may influence soils in many ways. The direct presence of salts is one aspect; changes in exchangeable cations on the soil colloids is another; and a third includes the indirect affects of salts on soil microbes, plant root activities, and the physical properties of soil colloids.[6]

---

[5] A bolson is a basin of interior drainage in an arid or semiarid region, whose floor is tending to be filled by a number of *alluvial fans* around its flanks.

[6] D. W. Thorne and L. F. Seatz, "Acid, Alkaline, Alkali, and Saline Soil," *Chemistry of the Soil*, ed. F. E. Bear (New York: Reinhold Publishing Corp., 1955), p. 240.

Some of the older soils in the arid regime show well-formed argillic horizons, while more recent surfaces generally lack such development. This feature perhaps may be a product of past climates of greater rainfall. The argillic horizon is an illuvial subsurface horizon in which silicate clays have accumulated. The genesis of this clay-enriched zone in arid climates has been explained by two diverse schools of thought. Nikiforoff[7] suggested the *in situ* formation of clay to produce a B horizon; for example, due to the subsurface layer remaining moist for longer periods than the surface horizon, chemical weathering is more active and clays are formed in place. Eventually this results in a texture difference between the A and B horizons, with the B horizon dominated by clay. A different viewpoint is discussed by Buol.[8] He notes that the clay of the B horizon may have resulted from illuviation even though there is no evidence of clay migration in the form of clay skins (*cutans*) in the profile.[9] In this particular environment it is thought that clay skins may have been destroyed by the repeated shrinking and swelling associated with wetting and drying.

A unique feature of many Aridisols is a residual surface layer of rock and gravel that is produced by deflation of fine particulate matter and is known as *desert pavement*. The remaining exposed rock and stone surfaces commonly have a thin black or dark brown stain from iron and manganese oxides, referred to as *desert varnish*.

Differentiation of Aridisols at the suborder level is based on the degree to which they have been weathered (Figure 7.5). The *Argids* (L. *argilla*, white clay) are normally found on older surfaces and contain an argillic or natric horizon; these same features are not found in the more youthful *Orthids* (Gr. *orthos*, true).

LAND UTILIZATION AND MANAGEMENT PROBLEMS

The main land uses associated with soils of arid climates revolve around either grazing or the intensive production of crops under irrigation in oasis settlements. Grazing predominates over most of the world's Aridisols. Grazing operations, however, frequently will require some degree of surface irrigation for the production of supplemental feed and for winter grazing. Thus, these two primary land-use functions are not always mutually exclusive of one another.

[7] C. C. Nikiforoff, "General Trends of Desert Type of Soil Formation," *Soil Science*, 43, 105-31.
[8] S. W. Buol, "Present Soil-Forming Factors and Processes in Arid and Semiarid Regions," *Soil Science*, 99, 45-48.
[9] Cutans are coatings of clay on the surface of soil peds and mineral grains, and in soil pores.

ARIDISOLS

Mineral soils that possess an ochric epipedon and one or more of the following properties: an argillic, calcic, cambic, petrocalcic, gypsic, natric or salic horizon, or a duripan.

ARGIDS

Aridisols that have an argillic or natric horizon.

Durargids have a duripan below an argillic horizon or a natric horizon within 40″ of the surface.

Haplargids have no duripan within 40″ of the surface and have an argillic horizon with less than 35% clay.

Nadurargids have a natric horizon with columnar structure overlying a duripan within 40″ of the surface.

Natrargids have a natric horizon and no petrocalcic horizon reaching with 40″ (1 meter) of the surface.

Paleargids have either a petrocalcic horizon with an upper boundary within 40″ of the surface or have an argillic horizon with more than 35% clay.

ORTHIDS

Aridisols that lack an argillic or natric horizon.

Calciorthids have a calcic or gypsic horizon with an upper boundary within 40″ of the surface.

Durothids have a duripan with its upper boundary within 40″ of the surface.

Paleorthids have a petrocalcic horizon with an upper boundary within 40″ of the surface.

Salorthids have a salic horizon within 30″ of the surface.

Figure 7.5   Suborders and Great Groups of the soil order Aridisol.

Grazing activities range from simple nomadic herding in the world's more economically depressed regions to complex large-scale commercial ranching in economically advanced areas (Figure 7.6). The soils are generally low in organic matter and nitrogen content; the amount reflects to a large degree the intensity of the regional moisture deficiency and the density of the native vegetation. Desert shrub and widely spaced bunch grasses make up the principal

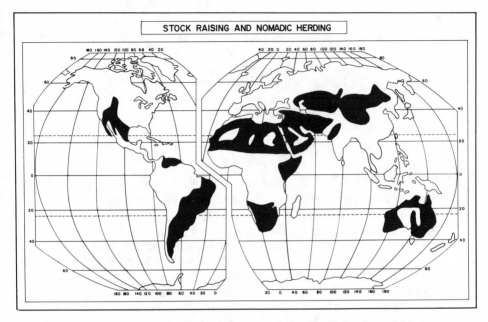

Figure 7.6 World distribution of stock raising and nomadic herding activities.

vegetative cover for animal consumption. The *carrying capacity*[10] of this vegetation varies greatly. In some areas of the southwestern United States it takes as much as 100 acres to supply forage for just one steer. As a consequence, extensive acreage is required to take care of even a moderately sized herd.

Animals grazing on desert vegetation can lead to a serious deterioration of the environment, unless conscientious management practices accompany range use. Overgrazing can speed up erosion and decrease soil permeability. The plant cover is removed, soils become compacted, and very loose or very wet soils actually may be displaced. This may be condusive to the rapid removal of surface water by overland flow, can cause destructive flooding, and can decrease soil moisture storage. Management objectives for the protection of range soils have been succinctly summarized by Wasser, Ellison, and Wagner:

> The basic objective of range management—a dense and well-dispersed cover of vegetation and mulch that provides adequate protection and imparts desirable characteristics to the

[10] Carrying capacity refers to the number of animals a given area of land can support with its natural vegetation.

soil—is economically attainable on most ranges by managing the grazing.

Soil management is accomplished on perhaps as much as 90 percent of the range area almost solely by manipulating grazing animals in accordance with four rules of good range usage:

Distributing livestock evenly to insure uniform use of forage on usable portions of each range unit; grazing the kind or kinds of animals that will economically utilize and perpetuate the desirable forage plants growing naturally on each range; adjusting the grazing use of each unit seasonally to meet the growth requirements of the desired forage plants; adjusting the numbers of grazing animals to attain an intensity of forage use that will maintain normal forage and soil productivity.

These considerations are basic to economical livestock production and to the maintenance of forage and soil productivity on all ranges. We call them the cardinal principles of range management.[11]

Aridisols that are placed under irrigation can be highly productive. Existing in areas with a high incidence of clear skies and high potential photosynthesis, the crop yields can be substantially above those in humid climates if an adequate supply of water is available.

The arid land farmer must confront problems other than scarcity of water that are unique to his environment. Saline accumulations in the soil may be at levels that are toxic to plants. The first requirement in reclaiming such land is to install drainage facilities. Subsequent applications of surface water then can remove harmful salts through leaching via subsurface drainage. If the soil has a high percentage of exchangeable sodium, additional treatment with gypsum or sulfur may be necessary. Gypsum provides soluble calcium that can replace sodium absorbed on clay particles. Sulfur applications serve the same basic purpose. Upon oxidation by soil bacteria, sulfuric acid is produced, which reacts with insoluble calcium carbonate in the soil to produce gypsum.[12]

In addition to these difficulties, the farmer may have to level the land, which can expose petrocalcic horizons, and he must also

[11]C. H. Wasser, Ellison, and R. E. Wagner, "Soil Management on Ranges," *The 1957 Yearbook of Agriculture: Soil* (Washington, D.C.: U.S. Govt. Ptg. Office, 1957), p. 636.

[12] W. H. Fuller and H. E. Ray, "Gypsum and Sulfur-bearing Amendments," *Agricultural Experiment Station Bulletin A-27* (Tucson: University of Arizona, 1963), pp. 3-11.

deal with the nutrient needs of the crops he plans to cultivate. Thorne notes that in the irrigated areas of the southwestern United States: "Nitrogen deficiencies are general. Phosphate deficiencies are frequent. Iron deficiency, or lime-induced chlorosis, is common with susceptible plants. Zinc, manganese, and boron deficiencies occur in restricted areas and especially with fruit crops."[13]

As if all these management problems, known to every inhabitant of the arid regions, were not enough, farmers must also battle the effects of turbulent winds and precipitation that often arrive in heavy downpours.

[13] Wynne Thorne, "The Grazing Irrigated Region," *The 1957 Yearbook of Agriculture: Soil* (Washington, D.C.: U.S. Govt. Ptg. Office, 1957), pp. 481-94.

Mollisols

8

Mollisols are mineral soils that have either a mollic epipedon or its equivalent.[1] A *mollic epipedon* (L. *mollis*, soft) is a surface mineral layer.[2] It possesses the following properties: (1) the surface retains a soft character, upon drying never becomes massive and hard; (2) base saturation is high ($\geqslant$ 50 percent); and (3) it is dark colored as a result of abundant humus. These relatively fertile soils are thought to be formed primarily by the underground decomposition of roots and surface organic residues (taken underground by animals) in the presence of bivalent cations, particularly calcium.

Most Mollisols are found in intermediate regions between the arid and humid climates.[3] As a result of their location they are known as transition zones, based on moisture availability that increases in the direction of the humid regimes. The area's vegetation indicates this intraregional moisture variation. Although dominated by a surface cover of grasses, the complex ranges from short, widely spaced bunchgrass along the arid margins to the richer and more

---

[1] In certain instances the surface horizon may meet all the specifications for a mollic epipedon except thickness. If this surface horizon lies above an albic layer that is superimposed on an argillic or natric horizon possessing the characteristics of the mollic layer, and if the combined thickness of the mollic layer and the epipedon meet the thickness requirements, the soil is classified as a Mollisol. (See Chapter 9 for a detailed explanation of an albic horizon.)

[2] Although the mollic epipedon is normally a surface horizon, in some very wet soils it may be overlain by an organic layer.

[3] Mollisol boundaries are irregular, with patches and interfingerings into both arid and humid realms in response to unique environmental conditions.

luxurient tallgrass prairie near the humid boundary. As might be supposed, the soil will reflect these differences in both moisture status and organic matter availability.

Associated primarily with semiarid and subhumid climatic regions, some of the most extensive contiguous stretches of Mollisols are found in: (1) the North American Great Plains, (2) from the western side of the Black Sea eastward beyond Lake Balkhash, (3) Manchuria, and (4) the Argentine Pampa (Figure 8.1).

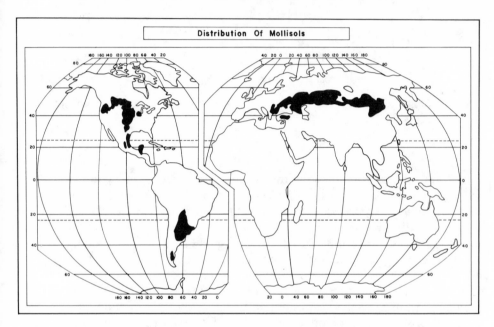

Figure 8.1   World distribution of Mollisols.

CLIMATE AND VEGETATION

The grassland environment in which Mollisols are normally formed experiences an annual moisture deficiency and is subject to highly variable precipitation. In certain years rainfall may exceed PE, permitting the expansion of the humid realm and a decrease in the dominance of semiarid/subhumid climates. In other years the reverse is true (Figure 8.2). Precipitation is more reliable than in the arid zone, but it is still highly erratic. The threat of drought is a continuing reality to the grassland farmer, and heavy downpours of rain accompanied by hail or tornadoes are not uncommon. An example of such an area's normal moisture supply and demand relationship is shown in Figure 8.3.

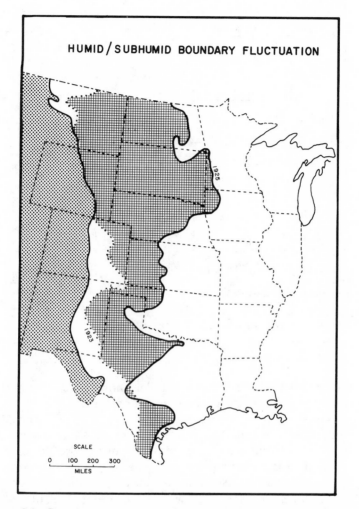

Figure 8.2 Fluctuations in the humid/subhumid boundary between the years 1923 and 1925.

Available energy for evapotranspiration is largely dictated by latitude and a continental location. Because these areas are in midlatitudes, they experience pronounced seasonal contrasts in receiving solar radiation. In addition, the low specific heat traits of continental interiors permit atmospheric temperatures to respond rapidly to radiation variation.[4] In many ways, the thermal properties

---

[4] Specific heat refers to the amount of energy required to produce a given temperature increase in a specified substance; for example, it takes approximately 580 cal/cm^3 to raise the temperature of water 1°C. It is estimated that the specific heat of water is about five times as great as that for the materials of the continents.

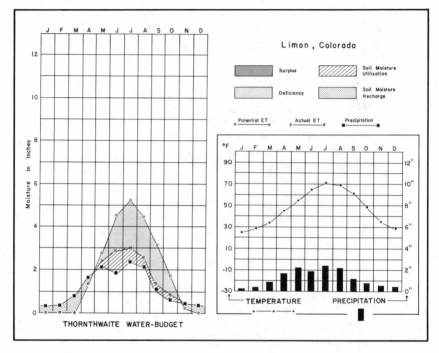

Figure 8.3   Water budget: Limon, Colorado (based on normal data).

of the grassland environment parallel those of the arid climates, except perhaps for absolute temperature maxima.

A grassland plant community in association with the Mollisols does not bear a simple relationship to mean annual precipitation. Rather, the distribution of soil moisture in the profile seems to be the crucial factor governing the presence of grasses. When the upper layers of the soil are moist during a considerable part of the year, yet the deeper layers are too dry for such deep-rooted plants as trees, a grassland forms and may dominate. Deep sandy soils that allow moisture to percolate deeply may, on the other hand, give support to trees in a region that is normally grass covered. Thus, the distribution of grasses is complex. They will, in general, withstand greater environmental extremes than forests—not only greater aridity, but also greater cold—and they will grow in wetter places.[5]

The character of the Mollisol profile is determined as much by the vegetation as it is by the climate. The softness of the epipedon, its coloration, and its available nutrients are all intimately related to the rooting and feeding habits and the life cycle of grasses. Variations

[5] C.P. Barnes, "Environment of Natural Grassland," *The 1948 Yearbook of Agriculture: Grass* (Washington, D.C.: U.S. Govt. Ptg. Office, 1948), p. 47.

in grassland plants bring about a wide variety of soils. In North America alone a minimum of seven grassland associations have been recognized.[6] However, there are two major grassland types—the true (tallgrass) prairie and the mixed (shortgrass) prairie, often called a *steppe.*

Tallgrass prairies exist in subhumid and humid lands. They are found primarily where soil moisture is deeper than in shortgrass lands but not deep enough to support trees. Their development is most extensive on the drier margins of the midlatitude forests. Smaller units exist within the boundaries of humid areas where calcareous parent material is present or where the ground water table is exceptionally high. The second type accounts for only a small percentage of the total area in which such vegetation occurs. Approximately 42 percent of all Mollisols have a cover of tallgrass prairie; the remainder is primarily shortgrass (steppe) vegetation, with forest occupying a small percentage.

Shortgrass vegetation occurs in drier semiarid areas, normally adjacent to deserts, where rainfall is sufficient only to moisten the upper soil layers during the warm season and to provide just enough support for shallow-rooted shortgrasses to survive. As aridity increases, the luxuriousness of the grasses decreases, and individual plants tend to be more widely spaced and of lower stature than their counterparts in more humid areas.

PEDOGENESIS

The most striking feature of the Mollisols, and the singular item of taxonomic importance at the order level, is their dark-colored epipedon. This is a product of the unique combination of the rooting habits, nutrient storing capacity, and the growth and decay cycles of grass coupled with the area's moisture regime.

Grasslands differ from forests in both the total amount of organic matter incorporated within the solum and in its distribution throughout the profile. In forested areas, the major contributing sources of soil humus are derived from leaf fall that accumulates on the surface. Percolation of rainfall through this decaying mass, or mechanical mixing by soil fauna, results in the organic debris being incorporated in the profile. Grasses also produce an organic mat as a result of the accumulation of their decaying aerial parts. In contrast to the rooting structure of trees, however, the dense, fibrous masses of grass roots prolifically permeate and uniformly distribute them-

[6] John E. Weaver and F. E. Clements, *Plant Ecology* (New York: McGraw-Hill Book Co., 1938), pp. 1-40.

selves throughout the profile. As the plants die and decay in an almost continuous process, a large quantity of humus is produced.[7] Thorp estimates that even relatively long-lived grasses (such as Bluestem) annually contribute between 520 and 900 pounds of raw organic matter per acre to the profile,[8] and in Wisconsin, under tall prairie, totals of 150 tons per acre have been recorded.[9] Although most of the grass roots and organic matter accumulate in the upper two feet of the solum, some grasses (such as *Andropogan gerardi*) extend to depths in excess of 80". In certain areas, the profile may also exhibit a subsoil of secondary organic matter enrichment, a result of both eluviation of the A2 horizon and root extension and proliferation during periods of drought (when the A horizon is depleted of moisture).[10]

Even though the total amount of organic material in the grassland profile is relatively high, the actual content in the epipedon varies considerably within regions. Along arid borderlands where moisture availability is low and where vegetation is widely spaced and short in stature, the minimum of 1 percent is approached; whereas totals of 8 to 10 percent are common in more moist regions, with even higher amounts in truly humid climates. The degree of *melanization*—darkening of the epipedon by organic matter—is reflected by darker hues of brown and an increased nitrogen content as soil organic matter increases. Allison states that under virgin conditions in the forested southeastern United States, soils averaged about 4,000 pounds of nitrogen per acre; whereas under a Boroll (Chernozem) amounts of 16,000 pounds per acre were recorded.[11]

Grasses, in general, require greater quantities of mineral nutrients—particularly calcium—than do trees. As a result, their organic remains are richer in base nutrients, which upon mineralization are subsequently returned to the soil. These released bases are again available for reuse by plants in a continuing cycle that under natural conditions maintains a relatively high degree of base saturation.

The structure of the mollic epipedon is granular (crumb) in those soils that have a texture coarser than silty clay loam, and is

[7] S. R. Eyre, *Vegetation and Soils: A World Picture* (Chicago: Aldine Publishing Company, 1968), p. 109.

[8] James Thorp, "How Soils Develop Under Grass," *The 1948 Yearbook of Agriculture: Grass* (Washington: U.S. Govt. Ptg. Office, 1948), p. 62.

[9] Francis D. Hole and Gerald A. Nielson, "Soil Genesis Under Prairie," *Proceedings of the Symposium on Prairie and Prairie Restoration* (Galesburg, Ill.: Knox College), p. 29.

[10] Ibid., p. 32.

[11] Franklin E. Allison, "Nitrogen and Soil Fertility," *The 1957 Yearbook of Agriculture: Soil* (Washington, D.C.: U.S. Govt. Ptg. Office, 1957), p. 90.

fine subangular blocky in textures that are finer. Hole and Nielson describe the processes involved in structure formation in the A1 horizon as follows:

> Wet-dry, shrink-swell, freeze-thaw and root expansion-root decomposition cycles in clayey A1 horizons foster the blocky structure. Prairie vegetation pumps available water out of a soil to a depth of several feet (a meter or two). Granular structure is formed by the passage of soil material through earthworms; by microbial production of soil-binding gums and other materials from biotic tissues; by formation of sesquioxide colloidal masses through weathering of primary minerals and mineralization of organic matter; by coagulation of humates by carbonates; by formation of binding organo-clay complexes; by knitting together of soil particles and aggregates by interlacing roots, some of which pierce the peds.[12]

The character of the eluviation-illuviation products within Mollisols can be explained largely in terms of the "materials being weathered" and the "weathering environment." Several writers have stressed that grassland soils do not normally form from hard rocks such as granite, but rather from unconsolidated calcareous sediments, with loess being the most extensive initial material worldwide.[13] Under tallgrass prairie, leaching of the soil column eventually results in a removal of carbonates and bases relative to their original content. Consequently, the soil column loses mass and experiences a reduction in pH. In the more extensive dry areas where potential evapotranspiration well exceeds precipitation, moisture rarely percolates beyond the root zone. Only the infrequent wet periods leach the very soluble salts of sodium and potassium from the profile. The less soluble calcium salts, in particular calcium carbonate ($CaCO_3$), have a tendency to accumulate in lower horizons because normal rainfall reaches only a few feet into the subsoil, where it gradually evaporates into the soil atmosphere or is utilized by plants. Under such dehydrating conditions calcium carbonate is carried into the subsoil in solution and later precipitated into hard nodules as the soil solution becomes supersaturated with mineral concentrates. A continuation of this process may eventually lead to the development of a petrocalcic horizon, similar to those produced in many Aridisols. It is evident that the drier Mollisols share pedogenic processes parallel to those dominating the Aridisol regime. A description of a Mollisol of the Typic subgroup of Paleustolls, and located in Cottle County, Texas follows:

---

[12]Hole and Nielson, *Prairie Restoration.*
[13]Thorp, *Grass*, pp. 55-56; Hole and Nielson, *Prairie Restoration*, pp. 28-34; Weaver and Clements, *Plant Ecology.*

Horizon	Depth (Inches)	
Ap	0-6"	Brown (7.5YR 4/4) silty clay loam, dark brown (7.5YR 3/3) moist and crushed; weak medium subangular blocky structure; very hard, firm; common roots; few strongly cemented $CaCO_3$ concretions up to 6mm in diameter; few siliceous pebbles and cobblestones on the surface and in the soil; mainly noncalcareous in matrix but weakly effervescent surrounding $CaCO_3$ concretions; abrupt smooth boundary.
B21t	6-14"	Dark reddish brown (5YR 3/2) silty clay, (5YR 3/3) moist and crushed; weak coarse prismatic structure parting to moderate medium and fine angular blocky structure; very hard, very firm; few roots; few strongly cemented $CaCO_3$ concretions as much as 6mm in diameter; few siliceous pebbles; cracks up to 3cm wide extend through lower boundary; noncalcareous in matrix; clear smooth boundary.
B22t	14-24"	Reddish brown (5YR 4/4) silty clay, dark reddish brown (5YR 3/4) moist; strong coarse prismatic structure parting to strong medium and fine angular blocky structure; extremely hard, very firm; few fine roots; few strongly and weakly cemented $CaCO_3$ concretions; few siliceous pebbles; cracks up to 2cm wide extend through lower boundary
B23t	24-38"	Reddish brown (5YR 4/4) clay, dark reddish brown (5YR 3/4) moist; strong coarse prismatic structure parting to strong coarse and medium angular blocky structure; extremely hard, very firm; few fine roots, mainly between peds; few strongly and weakly cemented $CaCO_3$ concretions; few siliceous pebbles; cracks up to 2cm wide; calcareous in matrix; gradual smooth boundary.
B24t	38-50"	Red (2.5YR 4/6) clay, dark red (2.5YR 3/6) moist on ped faces; streaks of dark reddish brown on faces of peds; interiors of peds are red (2.5Y 4/6) moist; moderate corase prismatic structure parting to moderate coarse and medium angular blockly structure; extremely hard, very firm; few fine roots; few strongly and weakly cemented $CaCO_3$ concretions; few siliceous pebbles; cracks up to 2cm wide extend to lower boundary; calcareous in matrix; clear wavy boundary.
B25tca	50-60"	Red (2.5YR 5/6) clay, red (2.5YR 4/6) moist, red (5YR 4/6) $CaCO_3$ coatings and dark reddish brown (5YR 3/2) clay coatings on faces of peds; moderate medium and coarse blocky structure; extremely hard, very firm; stringers or chimneys or yellowish red (5YR 5/6) moist $CaCO_3$ up to 15cm apart and 1cm in diameter; cracks extend to tops of the stringers of $CaCO_3$; estimated 5 percent of visable powdery $CaCO_3$ and few strongly cemented concretions; few dark pellets; few siliceous pebbles; calcareous in matrix; gradual wavy boundary. calcareous in matrix; gradual wavy boundary.

Horizon	Depth (Inches)	
B26tca	60-70″	Dark red (2.5Y 3/6) gravelly clay, light red (2.5Y 6/6) coatings 1-2mm thick of $CaCO_3$; moderate coarse angular blocky structure; very hard, very firm; few fine roots, more piping of $CaCO_3$ in form of powdery (5YR 4/6) bodies than in horizon above; about 15 percent of strongly cemented $CaCO_3$ concretions make up most of the gravel; few siliceous pebbles; few calcareous cobblestones; few dark pellets; few gypsum crystals; calcareous in matrix; abrupt wavy boundary.
B27tca	70-78″	Dark reddish brown (2.5YR 3/4) clay, black discontinuous coatings on faces of peds; moderate medium angular blocky structure; extremely hard, very firm; a layer high in coarse fragments contains carbonate pebbles up to 3 inches in diameter, 15 percent siliceous pebbles, and a few cobblestones of both carbonate and siliceous rocks; few gypsum crystals; calcareous in matrix; abrupt wavy boundary.
IIB3ca	78-82″	Dark reddish brown (2.5Y 3/4) clay, yellowish red (5 YR 4/6) moist; calcareous coatings 1-2mm thick; very fine and fine mottles of olive gray; moderate medium subangular blocky structure; extremely hard, very firm; few fine roots; few gypsum crystals; abrupt wavy boundary; few tongues extend into IIC horizon.
IIC	82-90″	Variegated grayish green (5GY 5/1) moist and reddish brown (2.5YR 3/4) moist; light clay; reddish yellow stains and thin seams of $CaCO_3$; clay coats on some faces; weak platy or blocky structure, retains part of apparent original rock structure; few gypsum crystals; noncalcareous in matrix.

The basic classification requirement for all Mollisols is the presence of a mollic epipedon. This criterion alone, however, is not sufficient. Certain soils with mollic epipedons may be classified in another order if they have other diagnostic features of greater taxonomic significance—perhaps an oxic horizon. (See Chapter XI for explanation of an oxic horizon.) There are seven recognized suborders of Mollisols (Figure 8.4): the *Albolls* (L. *albus*, white) are normally associated with Spodosols and possess an albic horizon (see Chapter 9); *Aquolls* (L. *aqua*, water) are found in areas of poor drainage and have characteristics related to wetness; the cold grasslands foster the development of *Borolls* (Gr. *boreas*, northern); *Udolls* (L. *udus*, humid), *Ustolls* (L. *ustus*, burnt), and *Xerolls* (Gr. *xeros*, dry), are distinguised from one another on the basis of soil moisture. Udolls have the least number of consecutive dry days, Xerolls always have over 60 consecutive dry days annually, and

Ustolls are intermediate in dryness. The last suborder *Rendolls* (Rendzina) does not exhibit characteristics typical of long-term weathering; rather the solum is developed from relatively recently exposed calcareous materials.

LAND UTILIZATION AND MANAGEMENT PROBLEMS

The high base saturation of the Mollisols and the semiarid to subhumid climate conditions that encouraged the growth of a natural cover of grass[14] have been conducive to large-scale commercial grain farming and livestock grazing (especially along the desert margins) or a combination of the two. James, in commenting on the occupance of the grasslands, states:

> In all the great grassland regions of the world the experience of man in his attempt to form permanent fixed settlements has been similar. In a general way there are the same problems of securing water, of clearing the grass sod, of combating insect pests, of building houses to withstand the extremes of weather in a land without the shelter of trees or hills, and of finding a suitable form of economy to make the settlement workable.[15]

Interestingly, these relatively fertile soils were once avoided by farmers in many parts of the world in preference for forested land. A major reason was the difficulty in turning over the grassland topsoil. The wooden plow, in use at the time, was incapable of cutting the dense, sod mat of grass roots. Consequently, it was easier for the farmer to fell trees and clear forest land for crops. With the introduction of the steel plow, slicing and turning over the sod was no longer such a problem. The steel plow, combined with increased demands for foodstuffs and greater efficiency in transporting agricultural commodities to market, fostered the rapid agricultural settlement in most of the world's middle latitude grasslands in the latter half of the nineteenth century.

The grains that are grown extensively on the Mollisols include wheat, rye, barley, corn, and sorghum, each having areal limitations based on unique habitat requirements. Wheat needs a cool, moist

---

[14]In large expanses of the grasslands there is sufficient evidence to indicate that early man, in controlling vegetation through the use of fire, has modified the natural vegetation from a forest cover to one of grass.

[15]Preston E. James, *A Geography of Man*, 2nd ed. (Boston: Ginn and Company, 1959), p. 352.

MOLLISOLS

Mineral soils that lack an oxic horizon and that have a mollic epipedon or its equivalent.

ALBOLLS

Mollisols that have an albic horizon beneath the mollic epipedon.

Argialbolls have an argillic horizon.

Natralbolls have a natric horizon.

AQUOLLS

Mollisols that are saturated with water at some period during the year, and have characteristics associated with wetness.

Argiaquolls have an argillic horizon.

Calciaquolls have a calcic horizon within 16" of the surface.

Cryaquolls are cold Aquolls of high latitudes or altitudes.

Duraquolls have a duripan within 40" of the surface.

Haplaquolls have a black epipedon.

Natraquolls have a natric horizon.

BOROLLS

Mollisols that are cool or cold and relatively freely drained.

Argiborolls have an argillic horizon.

Calciborolls have a calcic or gypsic horizon within 40" of the surface.

Cryoborolls are found in cold areas of the high latitudes or altitudes.

Haploborolls have a cambic or a blocky or prismatic subhorizon.

Natriborolls have a natric horizon.

Paleborolls have an argillic horizon at a depth > 24".

Vermiborolls have a mollic epipedon > 20" thick that shows strong evidence of worm casts and/or animal burrows.

RENDOLLS

Mollisols that have a layer more than 40% calcium carbonate below the mollic epipedon.

UDOLLS

Mollisols that are not dry for as much as 90 cumulative or 60 consecutive days per year.

Argiudolls have an argillic horizon.

Hapludolls have a brownish cambic horizon.

Paleudolls have an argillic horizon that does not have a clay content decrease ≥ 20% within 60" of the surface and are reddish in color.

Vermudolls have been intensively mixed by earthworms and their predators.

USTOLLS

Mollisols that are dry for more than 90 cumulative days per year, but not dry for as much as 60 consecutive days.

Argiustolls have an argillic horizon.

Calciustolls have a calcic or gypsic horizon within 40" of the surface.

Durustolls have a duripan within 40" of the surface.

Haplustolls have a cambic horizon or only slightly altered parent materials below the mollic epipedon.

Natrustolls have a natric horizon.

Paleustolls have a petrocalcic horizon within 60" of the surface or an argillic horizon where the clay does not decrease by 20% within 60" of the surface.

Vermustolls have > 50% of the epipedon and > 25% of the underlying horizon composed of worm casts or filled animal burrows.

XEROLLS

Mollisols of Mediterranean climatic regimes.

Argixerolls have an argillic horizon.

Calcixerolls have a calcic, petrocalcic, or gypsic horizon within 60" of the surface.

Durixerolls have a duripan within 40" of the surface.

Haploxerolls have a cambic horizon or only slightly altered parent material below the mollic epipedon.

Natrixerolls have a natric horizon within 60" of the surface.

Palexerolls have a petrocalcic horizon within 60" of the surface or an argillic horizon where the clay does not decrease by 20" within 60" of the surface.

Figure 8.4 Suborders and Great Groups of the soil order Mollisol.

period during the stages of early growth, and a hot, dry harvest season. For the seed to ripen, 90 frost-free days and an average summer temperature in excess of 57°F are required. The moisture demands of the plant vary with the thermal regime. On the poleward margins greatest wheat growth occurs when precipitation is between 10 and 40 inches. This is the region of spring wheat, planted in the spring after the last frost and harvested in the late summer. Most of the world's wheat, however, is of the winter variety. It is planted in the fall, begins its growth prior to winter, and is harvested early the following summer. The precipitation limits for winter wheat generally range between 20 to 70 inches.

Even though wheat grows well in this area, the greatest wheat yields are actually produced in the moist, marine west coast climates. The much-desired hard wheats with their high protein content, however, do not thrive in these wetter climates.[16] The greatest advantages of the dry grasslands for wheat farming are the very large expanses of level land that are suitable for the use of machinery and the area's unsuitability to moisture-demanding crops, such as corn.

Corn is a dominant grain on the grasslands where precipitation is either above or very near evapotranspiration demands and where the average summer temperatures are above 66°F. Of tropical origin, corn does best under summer conditions that are hot and humid. Rye and barley are normally concentrated on the poleward margins of the grasslands because they tolerate lower temperatures. Grain sorghum is becoming an increasingly important crop on the Mollisols, especially in the southwestern United States. These drought-resisting grains were introduced into the country from the semiarid parts of Africa. They are used for "fodder and as a binder crop in strip and terrace cultivation. In dry years grain sorghums provide a partial crop for feed and cover even when wheat withers and dies. They have been known to grow on as little as ten inches of annual rainfall."[17]

The average size of an operating unit for commercial grain farming in the semiarid regime is generally larger than its counterpart in any other area. Farms range from over 1,000 acres in the United States to well over 3,000 acres in the Soviet Union.

Management problems for the semiarid land agriculturist include the unreliability of moisture resources, as well as having to maintain the favorable fertility and structure characteristics typical of the native Mollisol. Farmers who occupied these areas had to develop techniques distinctly different from those of the humid

---

[16]Ibid., pp. 338-39.

[17]John W. Alexander, *Economic Geography* (Englewood Cliffs, N.J.: Prentice-Hall, Inc., 1964), p. 299.

lands from which many had emigrated. They soon found that the region's meager precipitation would not permit the continuous cultivation of crops, and they adjusted to a land-use pattern called *dry farming*. This technique allows the land to lie fallow for a year, permitting moisture to accumulate and to be stored for the production of a crop the following year. The cropland that is left bare is carefully cleared of vegetation, plowed, and harrowed in such a way to create a dust mulch on the surface, thereby reducing evaporation.

The most common problem facing the agriculturist of the grasslands, especially along its arid borders, is the continual threat of drought. Periods of deficient moisture frequently extend for several months, and occasionally several years may pass without any rainfall.

Sometimes the farmers of one county watch in despair as a heavy shower drenches a neighboring county. Sometimes, when the air is very dry, a shower forms, but the rain evaporates before it reaches the ground. From 1950 to 1956 there were large parts of the southern Great Plains that had no rain at all. The drought was so prolonged that even the cattle had to be moved away. With the droughts come the dust storms that blow the topsoil from vast areas of plowed land being used for dry farming.[18]

[18]James, *Geography of Man*, p. 342.

Spodosols

9

Spodosols are mineral soils that have either: (1) a *spodic horizon*, or (2) a *placic horizon* cemented by iron, overlying a fragipan,[1] and meeting all the requirements of a spodic horizon except thickness.

A spodic horizon is generally found beneath an eluvial mineral layer, which under virgin conditions is frequently a light-colored *albic horizon*.[2] The character of this diagnostic feature is that of a soil stratum in which "active" *amorphous* (without crystalline structure) materials composed of organic matter and aluminum, with or without iron, have precipitated.[3] A placic horizon, on the other hand, is a thin black to dark reddish pan that appears to be cemented primarily by iron.[4] This feature has a pronounced wavy or involuted

---

[1] A fragipan (modified from *L. fragilis*, brittle, and *pan*—meaning brittle-pan) is a subsurface horizon that is low in organic matter, has a high bulk density relative to the horizons above, and is seemingly cemented when dry, having hard or very hard consistence. Their genesis is obscure, but their hardness is largely attributed to the close packing and binding by clays, and their brittleness is thought due to weak cementation.

[2] An albic horizon is the one from which clay and free iron oxides have been removed, or in which the oxides have been segregated to the extent that the color of the horizon is determined by the color of the primary sand and silt particles rather than by coatings on these particles.

[3] The Soil Survey staff uses the term *active* in this situation to describe materials having a high exchange capacity, large surface area, and high water retention.

[4] A pan is a layer or horizon within a soil that is firmly compacted or is very rich in clay.

form, generally ranges from about $1''$ to $4''$ in thickness, and is normally found within the upper $20''$ (50cm) of the mineral soil.

The spodic and placic horizons are absent in the soils of arid regions, yet are found in humid areas ranging from the tropics to the tundra margins. They are best developed and show maximum areal extent in humid regimes that are cold and forested. These conditions are dominantly found in the northern portions of North America and Eurasia (Figure 9.1). Due to the absence of extensive land masses in the middle latitudes of the southern hemisphere, no parallel extensive region of Spodosol development occurs south of the equator.

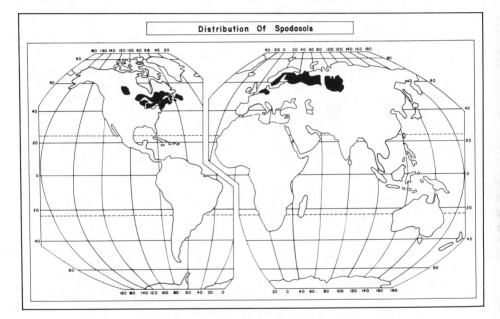

Figure 9.1 World distribution of Spodosols.

CLIMATE AND NATIVE VEGETATION

The most extensive continuous stretches of Spodosols occur in areas of subarctic climate. In this region winters are long and cold, there is great seasonal variation in air temperature, and precipitation is concentrated in the summer.

The subarctic region experiences the earth's greatest temperature ranges. Verkhoyansk in the Soviet Union, for example, has a mean monthly temperature variation from approximately -57°F in

January to about 58°F in July—a range of 115°F. A similar, although not as intensive, pattern can be seen in Figure 9.2. Summers are short, with normally less than four months having temperatures above 50°F. Warm spells may raise daily temperature maxima into the 80s, but at the same time frost may occur on any given day. Winter temperatures get very low, being lower only over permanently ice-covered areas such as Antarctica and Greenland.

Precipitation in the subarctic is relatively light. However, the regionally low temperatures result in low evapotranspiration demands, so the surface, amply supplied with moisture, experiences an annual surplus.

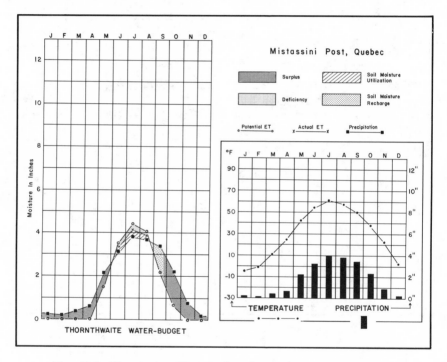

Figure 9.2 Water budget: Mistassini Post, Quebec (based on normal data).

Because they can endure a strenuous winter season and adapt to a brief summer, the evergreens dominate in the subarctic region and make up the most extensive forest formation (often called the *boreal forest*) on the earth's surface. These plants, with their xeric (low moisture) adapted needle leaves, successfully outcompete broad-leaved trees in this environment, where water is unavailable during the part of the year that soils are frozen. The evergreens (as the name implies they do not shed their leaves at one time) have an advantage over deciduous trees, which shed their leaves during winter. Leaf-shedding trees must grow new sets of leaves before they can resume photosynthesis; thus they do not have maximum use of the short growing season.[5]

The boreal forest zone is comprised of two distinct plant formations—one in North America and the other in Eurasia. In the eastern boreal forest of North America, white spruce and balsam fir occupy well-drained sites, and black spruce, jack pine, larch, and tamarack dominate in areas of poor drainage. The Eurasian area has a fairly simple composition in western Europe, where pine and spruce dominate the landscape. Farther east other species begin to take over, especially the Siberian fir, spruce, and larch, and the dwarf Siberian pine.

The main characteristic of the boreal forest is uniformity. Over extensive areas dense, homogeneous stands of evergreen conifers (cone-bearers, as most evergreens are) overshadow the surface and allow little sunlight to penetrate for the support of undergrowth. Therefore, only small herbaceous plants and fungi that require little direct sunlight can thrive. Under such shaded conditions, and relatively low temperatures over a significant portion of the year, microbial activity is low and surface organic debris is decomposed slowly. This results in a mineral soil surface covered with litter made of needle leaves and semidecomposed branches. Decomposition is limited to the degree that annual leaf falls contribute to an ever increasing mat of organic matter.

The deep shade provided by the forest canopy shields the ground from direct sunlight. Evaporation is thus reduced, and the soil normally remains in a moist to wet state, thereby encouraging leaching of bases from the profile. Most coniferous trees are not high in demands for soil nutrients and are capable of surviving on soils with a relatively low base status. Frequently they inhabit sites consisting of highly siliceous materials. This, of course, means that the organic products accumulating on the surface are also low in bases and will produce acidic humus as they decompose.

Spodosols do not form *only* under coniferous vegetation, however. Even in the subarctic, large areas of heath also show features that are characteristic of this soil order, and in warm climates spodic horizons exist under savanna, palms, and mixed forest. The occurrence of the spodic horizon even in some soils of the tropics has

---

[5] In certain poleward areas the winters are so severe that the presence of leaves is a disadvantage to the plants, and only the slower growing deciduous species, such as birch, aspen, and larch, are capable of surviving.

generated a controversy over its pedogenesis in warm areas. Some researchers feel that the tropical spodic horizon is not a product of present climatic and vegetation influences, but rather is associated with a past—and different—environment.

PEDOGENESIS

The Soil Survey staff notes that under optimal conditions a spodic horizon may form in as little as a few hundred years, indicating that pedogenic processes are highly active. The processes that influence the development of Spodosols in cool, moist climates are known collectively as *podzolization*. The term is derived from the Russian word *podzol*, which is translated as "ash." This may seem a strange way to describe a soil or soil-forming processes, unless one remembers that the original use of the term by pedologists was to characterize the soil's color—ashen grey. In nineteenth-century Russia, the soils first classified as podzols had a surface mineral horizon from which most minerals, except silica, had been removed by leaching. Frequently this produced a sandy, bleached A horizon. The physical mixing of organic matter into this soil layer (for example, by burrowing animals or percolating water) subsequently gave a grayish color to the horizon. This light-colored layer is now called the albic horizon. Although an albic horizon is frequently found in Spodosols, its presence is not an absolute requirement for classifying a soil within this order. What is necessary, however, to produce the typical Spodosol morphology are the reactions involved in the movement and accumulation of sesquioxides and organic matter.

Before humus and sesquioxides can change their locations, exchangeable bases must first be displaced and leached from the upper part of the solum by hydrogen ions (H+). The ideal setting for such action is where the soil undergoes a through moistening of the profile (a flushing moisture regime), where temperatures are low enough to inhibit the microbial decomposition of organic matter, and where the flora has a very low demand for base nutrients. These conditions are best fulfilled in the subarctic regions where the vegetation consists primarily of coniferous forest with moss, lichens, and subshrubs, and where underbrush and grass are few in number or completely absent. The organic litter produced by these plants is low in calcium and nitrogen, and does not decompose rapidly. It has a high content of compounds such as lignin, wax, and resin, which are resistant to breakdown, and it may contain tannins and terpenes, which further inhibit decay. The character of the subarctic boreal region retards the biochemical recycling of base nutrients, and surface accumulation of forest litter results.

The precipitation infiltrating this organic litter leaches organic compounds, deficient in bases, that are of the fulvic acid type.[6] This acidic solution percolates through the solum, first removing a considerable portion of the exchangeable bases by hydrogen ion (H+) displacement. The rate of the base desaturation varies, depending largely on the parent materials. It is rapid on stable coarse-textured acid materials with a low base-exchange capacity, but slow on less stable materials where a greater concentration of calcareous or saline parent materials must be removed. Removing the free lime and/or salts makes the clay particles disperse. The dispersion (called peptization) speeds their vertical movement and accumulation in lower horizons. As a consequence, in time the solum develops a profile that has different textures and is chemically altered. The surface horizon is depleted of fine-sized particles, giving it a coarse texture, and the subsurface horizon is enriched with fine-sized clay and silt particles. Calcium and magnesium are washed out of the topsoil, leaving behind a layer of silicate minerals. After this initial stage has been reached, the soil solution—in the absence of bases—becomes more acid. Under this condition iron and aluminum may be liberated from their mineral bonds and migrate downward from the surface horizon. These *sesquioxides* (oxides with one and one-half oxygen atoms to every metallic atom) most frequently translocate, or move, in two forms—as *inorganic cations* and as *metal-organic complexes.*

Stobbe and Wright have summarized these concepts of sesquioxide translocation. Referring to early research on the development of Spodosols, when it was thought that "strong acidity created by the decomposition products of organic matter brought about the solution of sesquioxides (as inorganic cations), which then moved to lower levels where a higher reaction caused their precipitation,"[7] they say that the pH level of these soils is normally too high for the solubility of ferric iron. Given reducing conditions, however, where the solum is saturated with water for an appreciable portion of the year, ferrous iron may remain in solution and migrate down the profile as ferrous ions, subsequently oxidizing during periodic dryness. That situation may account for sesquioxide translocation as

---

[6] Fulvic acids are organic substances that will remain in solution after acidification of a dilute alkali extract taken from the soil.

[7] P. C. Stobbe and J. R. Wright, "Modern Concepts of the Genesis of Podzols," *Soil Science Society of America Proceedings,* 23 (1959), 162.

inorganic cations, at least in poorly drained and high water-table sites.

The metal-organic complexes are thought to develop from the interaction of organic acids with sesquioxides to form soluble organic-mineral complexes. These complexes migrate with the percolating soil solution to lower horizons. Subsequent conditions may lead to the precipitation of the sesquioxides within the illuvial horizon and the formation of a spodic horizon.

It is possible that:

1) microorganisms destroy the complex organic mineral compounds that initially mobilized the Fe and Al;
2) insufficiency of infiltrating precipitation fails to carry colloids and solutes farther down the profile and forces precipitation in the illuvial horizon, as dryness occurs;
3) sieving action by soil particles can lead to the clogging of pore spaces by colloids;
4) negative charges on the clay films immobilize positively charged Al and Fe ions; or
5) hydrolysis of the organic-metal complex is induced by changes in the pH level.[8]

The spodic horizon is most readily recognized by its color and structure. The upper boundary is abrupt, and hues, values, and chroma change markedly with depth within a few centimeters. The lowest color values, reddest hues or highest chromas, are always in the upper part of the horizon. Texture is most commonly sandy, coarse-loamy, or coarse-silty, with structure that is either absent or present as crumb, granular, platy, or weak blocky or prismatic. The spodic horizon must be at least 1″ thick and have a relatively high cation-exchange capacity. Commonly, it will also exhibit a secondary maximum of organic matter concentration. Since the spodic horizon is an illuvial layer and accumulates the minerals removed from the soil's epipedon, an impoverished mineral zone necessarily forms immediately above the horizon. This eluvial layer frequently has the "classical" feature of the originally identified podzol soil, or what is now called the albic horizon. If an albic horizon is present, it has

[8]*Soil Science Society of America Proceedings*, pp. 161-63; National Cooperative Soil Survey; *Soil Taxonomy* (Washington, D.C.: U.S. Govt. Ptg. Office 1970), pp. 3-19 to 3-24; S. W. Buol, F. D. Hole, and R. J. McCracken, *Soil Genesis and Classification* (Ames, Iowa: Iowa State University Press, 1973), pp. 256-57.

light color and relatively coarse texture. The coloration mainly comes from the siliceous sand and silt particles rather than from coatings of other minerals that migrate downward.

Following is a profile description of a Spodosol of the suborder Aquod. It is known as a Typic Sideraquod, and shows an appreciable accumulation of iron in the spodic horizon relative to the organic carbon. This soil, in Curry County, Oregon, formed in beach deposits under a vegetation of pine, bracken fern, grasses, and sedges. The mean annual precipitation is about 66″ coming mostly in winter. The soil is saturated in the winter months. The mean annual temperature is about 50.8°F, and the mean July temperature is about 59°F. Slopes are very gentle.

Horizon	Depth (Inches)	
A0	2½-0″	Very strongly acid muck with some needles, leaves and grass litter.
A1	0-4″	Very dark gray (10 YR 3/1) mucky fine sandy loam, dark gray (10yr 4/1) when dry; moderate, fine subangular blocky structure parting to moderate fine granular structure; soft, very friable, slightly plastic; many moderately coarse pores; many roots; very strongly acid, pH 5.0; gradual wavy boundary.
A2	4-18″	Gray (N 5/0 or 6/0) fine sandy loam, white (N 6/0 or 8/0) dry; massive; hard, firm, slightly plastic; few fine pores; many roots; strongly acid, pH 5.2; abrupt wavy boundary.
B21h	18-22″	Dark reddish brown (5YR 2/2) mucky loam, dark reddish gray (5YR 4/2) dry; weak medium subangular blocky structure or massive—some cementation; soft, very friable; many fine and moderately coarse pores; common roots; very strongly acid, pH 5.0; abrupt wavy boundary.
B22ir	22-32″	Yellowish brown (10YR 5/4 and 5/6) and dark reddish brown (5YR 3/4) loamy sand; common medium and coarse dark reddish brown (5YR 3/4) mottles; massive, very hard, very firm; few fine pores; medium acid, pH 5.4; clear wavy boundary.
B3	32-48″	Yellowish brown (10YR 5/4) loamy fine sand; few medium distinct strong brown (7.5YR 5/6) and few fine prominent dark reddish brown (5YR 3/4) mottles; massive; slightly hard, firm; many coarse pores; medium acid, pH 5.6; gradual wavy boundary.
C	≥48″	Pale yellow (2.5YR 8/4) loamy fine sand; yellowish brown mottles; massive slightly hard, loose; many coarse pores; reddish brown concretions; medium acid, pH 6.0.

Spodosols are divided into four suborders (Figure 9.3):

1. *Aquods*. These are wet soils with free iron in very small amounts. The brown and reddish brown colors of the humus in the spodic horizon tend to mask other colors. They range in location from the arctic margins to the equator.
2. *Ferrods*. Not a soil found in the United States, its features include a considerable accumulation of iron and a relatively high base-exchange capacity.
3. *Humods*. These soils are high in accumulated humus, with little or no iron.
4. *Orthods*. These are the most common Spodosols in the northern parts of Europe and North America, formed mainly in coarse, acid Pleistocene or Holocene topsoils. They usually exhibit organic, albic, and spodic horizons, with or without a fragipan.

LAND UTILIZATION AND MANAGEMENT PROBLEMS

The Spodosols are the life-supporting medium for the world's largest contiguous expanse of boreal forms of vegetation. However, trees of construction timber quality also thrive in the more southern regions of the zone. One such area was the famous "Lake Forest" of North America, which once extended from Minnesota to the coasts of New England. Only remnants remain of this magnificent forest, which at the time of colonization consisted primarily of white and red pine and eastern hemlock with heights exceeding 200 feet. The excellent character of the forest led to its early and rapid exploitation, a process which was fairly complete by the end of the nineteenth century.

After the loggers had removed the timber, extensive acreages of Spodosols were settled by farmers, who were most often unaware that these highly infertile soils could not support the type of agriculture that takes nutrients out of the soil without replacing them. The result was considerable despair to the farmers, whose production costs in conditioning these croplands were so high that general farming became prohibitive, and the land was frequently abandoned.

Within the boreal forest proper, extensive areas have been cleared by the demand for pulpwood and by fire. Many almost pure stands have been exploited on a large scale because the trees are desirable for making into newsprint. Besides the careless (in many

SPODOSOLS

Mineral soils that have a spodic horizon with an upper boundary within 2 meters of the surface or that have a placic horizon cemented by iron that rests on a fragipan or on an albic horizon that rests on a fragipan, and that meets all requirements of a spodic horizon except thickness.

AQUODS

Spodosols that have characteristics associated with wetness.

Cryaquods have a cryic temperature regime.

Duraquods have strongly cemented or indurated albic horizon.

Fragiaquods have a fragipan below the spodic horizon but lack a placic horizon above the fragipan.

Haplaquods have in > 50% of each pedon, a spodic horizon with some subhorizon in which the ratio of free iron to carbon is < 0.2.

Placaquods have a placic horizon that rests on a spodic horizon or on an albic horizon that is underlain by a spodic horizon.

Sideraquods are other aquods.

Tropaquods have a mean annual soil temperature of 8°C or higher and a mean annual range in soil temperature of less than 5°C.

FERRODS

Spodosols that have: (1) a spodic horizon that has in all subhorizons a ratio of percentage of free iron to percentage of carbon of 6 or more, and (2) do not have characteristics associated with wetness.

HUMODS

Spodosols that are relatively freely drained and that have large accumulations of organic carbon relative to iron in some or all subhorizons of the spodic horizon.

Cryohumods have a cryic temperature regime.

Fragihumods have a fragipan below the spodic horizon.

Haplohumods are other humods.

Placohumods have a placic horizon.

Tropohumods have an isomesic or warmer iso-temperature regime.

ORTHODS

Spodosols that are relatively freely drained and have horizons of aluminum, iron, and organic carbon accumulation.

Cryorthods have cryic or pergelic temperature regime.

Fragiorthods have a fragipan below the spodic horizon.

Haplorthods are other orthods.

Placorthods have a placic horizon.

Troporthods have a mesic or warmer iso-temperature regime.

Figure 9.3 Suborders and Great Groups of the soil order Spodosol.

cases) destruction of these magnificent forests, another concern is that these large clearings do not quickly revert to their former state after cutting. Instead, species of birch and aspen are the first to return. These deciduous trees normally produce large quantities of very light seeds capable of being carried tremendous distances by wind. In addition, the seeds can germinate in areas where they are exposed to the dessicating (drying) effects of the sun and wind—much more successfully than do the seeds of conifers. As a result, instead of deciduous trees occurring as isolated plants or in small clumps, as they did in the original forest, they now occupy extensive areas where clear-cutting or burning has taken place.[9]

Forestry has been the dominant economic activity on the Spodosols, but agriculture is of substantial importance as well. Cultivated crops, hay fields, and pasturelands are all becoming significant parts of the Spodosol landscape. Their successful cultivation in a demanding environment and on soils of inherently low fertility is the result of wise land-use practices and modern technology.

With surplus rainfall, the continued washing of the profile depletes the solum of calcium, magnesium, and nitrogen—essential nutrients for cereal grains and pasture grasses—and induces low base saturation, in some horizons less than 10 percent. This leaves a soil whose exchange complex is dominated by the hydrogen ion ($H+$) and which is acid throughout. Due largely to their coarse texture, Spodosols also normally have a low capacity to store water. To utilize these soils for agriculture their limitations must be recognized and *amendments* provided.[10] The earliest attempts to make these soils productive involved incorporating crushed marl and organic matter (in the form of animal manure and root residues) into the soil. The effect was twofold: (1) liberation of bases, especially calcium, from the marl reduced soil acidity. Increased pH had the effect of limiting the peptization of clays and encouraging their flocculation (forming into loose clusters). Hence, further translocation of clay was retarded and soil structure was improved; (2) Nitrogen content and soil water-holding capability increased as organic matter was incorporated within the plow-layer of the solum.

Marling is not practiced today; instead specially treated limestone is used as a conditioner on acidic soils. The limestone is first

[9] S. R. Eyre, *Vegetation and Soils* (Chicago: Aldine Publishing Company, 1971), pp. 47-55.

[10] An amendment is any material, such as lime, gypsum, sawdust, or synthetic conditioners, that is worked into the soil to make it more productive. Strictly, a fertilizer is also an amendment, but the term is most commonly used to refer to added materials other than fertilizers.

crushed, sized, and calcinated. Calcination involves heating the limestone at a temperature about 1600°F to burn off carbon dioxide. The resultant product, CaO, is known as quicklime. It will readily react with soil water to produce a calcium hydroxide (CaO + H_2O = $Ca(OH)_2$), which neutralizes soil acidity by replacing exchangeable hydrogen (H) and converting it by recombination into water (H_2O). The importance of limestone as a soil amendment in acidic soils is illustrated in Table 12, which shows the yield increase of crops provided with two tons of limestone per acre. These increases amount, in some cases, to as much as 33 percent.

Table 12   Yield Increases of Selected Crops after an Application of Two Tons Limestone per Six Year Rotation *

Crop	Yield/Increase per Acre
Corn (silage)	1.0 tons
Oats	5.8 bushels
Wheat	7.8 bushels
Hay	1.8 tons

*Data is for New York State.
Source: N.C. Brady, R. A. Struchtemeyer, and R. B. Musgrave, "The Northeast," The 1957 Yearbook of Agriculture: Soil (Washington: U.S. Govt. Ptg. Office, 1957), p. 604.

In addition to lime, individual plants have specific nutrient requirements, and the land manager who supplies them is the one to achieve maximum crop yields. Dairy farmers cultivating Spodosols in the United States, for example, usually have a rotation that includes small grains, hay, and pasture. The hay consists primarily of red clover, timothy, and alfalfa, all of which require lime and high potash fertilizers to maintain productive stands. These soils also tend to be low in phosphorus and nitrogen. Top dressings of nitrogen, phosphate, and potash mixtures have produced remarkable responses in growth and yields of hays and grains, and have increased pasture production by as much as 300 percent.

The Spodosols and their associated climate regions are well suited for root crops, especially potatoes and sugar beets. Potatoes are very susceptible to their environment. They grow well in acid soils with a pH of about 4.4 to 5.4. If the pH falls much below 5.0, soluble manganese increases in the soil and may become toxic for the potatoes. In that case it may be necessary to apply finely ground dolomitic limestone in sufficient quantity to raise the pH level and reduce the manganese. However, if the pH level is raised too high, the plants will experience scabbing. Potatoes also require the careful control of phosphorous, potassium, nitrogen, and copper. Boron, magnesium, and sulfur are deficiencies that also may have to be overcome if plants other than potatoes are grown in this area.[11]

Some of the world's best known nurseries and market gardens, such as those in the western Netherlands, southeast England, Denmark, and northern Germany occur on Spodosols. Many years of deep plowing, the use of fertilizers, and heavy manuring have created soils that are highly productive for human needs. These sandy soils lack the sticky character of clay soils, and their ease in cultivation has earned them the reputation of being "light." Since the soils are also highly permeable and contain relatively large pore spaces, plants can produce abundant fibrous roots with relative ease. This porous character can be both an advantage and a disadvantage, however. Sandy soils have a low capacity to store water; hence they warm up quickly in the spring and give plants an early start. On the other hand, drought periods have a rapid impact on vegetation that is sustained on sands. This frequently means the additional expense of supplemental irrigation for high value crops. The intensively utilized Spodosols require a considerable capital investment to insure maximum production, and they are profitable only where a market is available and the demand for produce is sufficient.[12]

[11]R. J. Muckernhirn and D. C. Berger, "The Northern Lake States," *The 1957 Yearbook of Agriculture: Soil* (Washington, D.C.: U.S. Govt. Ptg. Office, 1957), pp. 550-51.

[12]E. J. Russell, *The World of Soil* (London: Collins Clear-Type Press, 1957), pp. 238-43.

Alfisols and Ultisols

10

Alfisols and Ultisols are mineral soils of humid moisture areas. They share the common characteristic of an argillic (clay) horizon, yet differ appreciably in their base status. The Alfisols usually have moderate to high base saturation and an ochric epipedon. The Soil Survey staff recognizes six diagnostic epipedons: mollic, umbric, anthropic, plaggen, histic, and ochric. The first five have soil properties that meet specific quantifiable limits. Ochric epipedons, on the other hand, are surface horizons that lack sufficient development to be classified as one of the first five—they are either too light in color, too high in chroma, too low in organic matter, have too high an n value,[1] or are too thin to be mollic, umbric, anthropic, plaggen, or histic.

The Ultisols (L. *ultimus*, last) are more thoroughly weathered and have experienced greater mineral alteration than any other midlatitude soil. In addition, extensive leaching has resulted in low base status either in or immediately below the argillic horizon. Yet, the overall gross morphology and horizon sequence of the Ultisols and Alfisols are similar. This has led some researchers to believe that with time and further weathering Alfisols may eventually degenerate into Ultisols.[2]

[1] The n value refers to the ratio between the water percentage under field conditions and the percentage of inorganic clay and humus. The n value is helpful in predicting whether the soil may be grazed by livestock or support other loads, and the degree of subsidence that would occur following drainage.

[2] H. D. Foth and L. M. Turk, *Fundamentals of Soil Science*, 5th ed. (New York: John Wiley and Sons, Inc., 1972), p. 260.

Alfisols are found in the middle latitudes and the intertropical zone. In the middle latitudes they occur between the Mollisols of the semiarid and subhumid climates and the Spodosols and Ultisols of the humid regimes. Intertropical locations normally occur in a transitional area between the Aridisols of the desert and the Ultisols and Oxisols of the humid climates (Figure 10.1). The Ultisols show a very simple areal pattern, being confined to humid subtropical climates and some relatively youthful tropical landscapes (Figure 10.2).

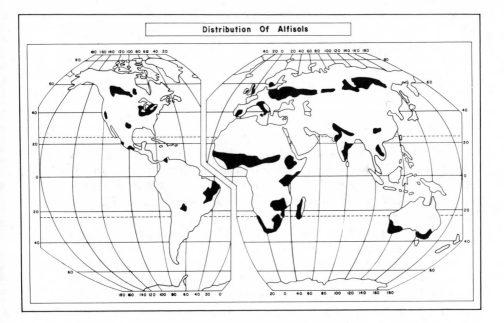

Figure 10.1   World distribution of Alfisols.

No other soil order with mature profile development, occurring on an extensive scale, exists in as many diverse climatic and vegetation environments as do the Alfisols. These soils are recognized in micro to macro thermal regimes, in moisture realms ranging from humid to seasonally arid, and under flora that varies from broadleaf deciduous trees to the thorn trees and tall savanna grass of the tropics. Figures 10.3 and 10.4 show two diverse environments in which Alfisol development is dominant. Each example depicts the moisture status of a recognized humid climate, yet each also shows a seasonal period during which some degree of soil desiccation is expected.

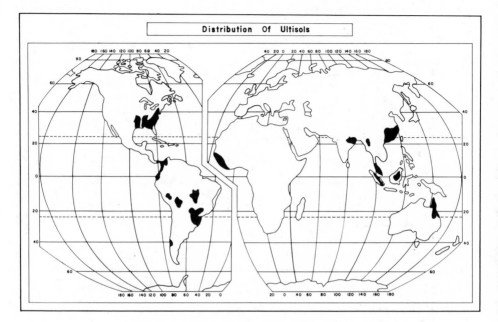

Figure 10.2 World distribution of Ultisols.

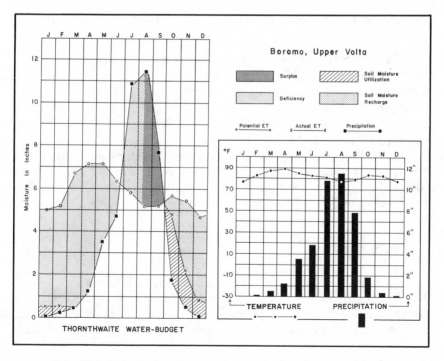

Figure 10.3 Water budget: Boromo, Upper Volta (based on normal data).

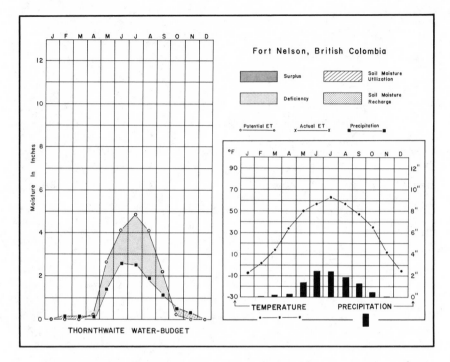

Figure 10.4   Water budget: Ft. Nelson, British Colombia (based on normal data).

The character of the two temperature traces varies considerably, with ranges from 11.5 F° in Boromo to 70.6 F° in Fort Nelson. The maximum and minimum mean monthly temperatures also differ considerably. Boromo has a temperature low of 77.9°F and a high of 89.4°F, while the monthly minimum in Fort Nelson is −8.4°F and the maximum is 62.2°F.

The Ultisols of the middle latitudes lie south of the Alfisols in a humid subtropical climate. In contrast to the environment of the Alfisols, winters are milder, the temperature regime is usually less continental, and precipitation is normally higher (Figure 10.5).

PEDOGENESIS

Unlike the soils of the moisture-deficient regions, where calcification is dominant, or the subarctic, where podzolization is active, most Alfisols and Ultisols occur in fairly dependable thermal and moisture regions that are not subject to major climatic

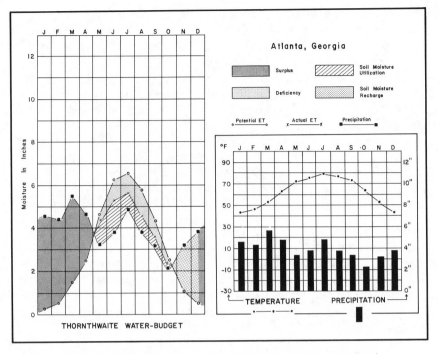

Figure 10.5 Water budget: Atlanta, Georgia (based on normal data).

extreme.[3] The characteristics of these soils result from the interaction of the "distance-modified processes" attributed to all the earth's major pedogenic regimes.

In the midlatitudes, these soils seem to be associated with relatively stable landscapes in which soil water has eluviated and illuviated clays more rapidly than erosion has truncated the surface horizon.[4] The argillic horizon (as we know, a subsurface horizon with an accumulation of translocated clay) and its associated base status, although present largely because of the action of percolating soil water, are also products of the long-term influences of vegetation and climate.[5]

[3] In addition to such situations as a high temperature range or deficient precipitation, uniformity of monthly temperature and/or precipitation also can be considered a climatic extreme.

[4] Guy D. Smith, "Lectures on Soil Classification," *Pedologie*, IV (Ghent, Belgium: State University of Ghent, 1965), 67.

[5] The following explanation of the character and genesis of the argillic horison is basically a summary of the Soil Survey Staff's discussion of this feature in *Soil Taxonomy* (Washington, D.C.: U.S. Govt. Ptg. Office, 1970), pp. 3-8 to 3-17.

The evidence of the presence of an argillic horizon is closely related to the processes that were instrumental in its formation, and specifically rests on the character and orientation of the transported clays. Since there is a strong similarity in the fine clays of both the eluvial and illuvial horizons, it is assumed that the clays migrate downward rather than form as decomposition products being synthesized into clays *in situ*, and since water is the agent moving the clay:

> the translocated clay tends to form coatings of oriented clay particles on the channels through or into which water moves or in which it stands. These channels are principally crevices bound by the cleavage faces of peds, and the pores left by roots or animals.[6]

When they are first moved and deposited, the clays tend to be oriented with their long axes parallel to the surface upon which they adhere, and they have the appearance of a surface coating smeared upon the ped faces.[7]

The formation of this diagnostic horizon requires certain pre-existing conditions:

1. either the parent material must contain very fine clays or subsequent weathering must be capable of producing them;
2. these very fine clays must be subject to dispersion. The presence of certain mineral salts, such as carbonates and free oxides, will tend to flocculate (cluster) clay particles, making it more difficult to displace them. In such cases, the chemical weathering and removal of these salts must precede clay migration.

A dry soil with these traits will, when wet, experience a disruption of its fabric and the subsequent dispersion of clay. At this point the eluvial clay can be physically carried by the soil solution to a depth where percolation ceases. This process is described below:

> Water percolating in noncapillary voids commonly is stopped by capillary withdrawal into the soil fabric. During this withdrawal the clay is believed to be deposited on the walls of the noncapillary voids. This would explain why illuvial clays are commonly plastered on the ped faces and on the walls of pores.

---

[6] Ibid., p. 3-11.

[7] The thin clay coatings have a variety of names, such as clay skins, clay films, clay flows, and illuviation cutans.

Such a mechanism for clay movement and deposition is favored in several ways by a seasonal moisture deficit; first, as mentioned above, wetting a dry soil favors the dispersion of the clay; second, on drying, cracks form in which percolation of gravitational water or water held with low tension can take place; third, the halting of percolating water by capillary withdrawal is favored by the strong tendency for dry soil to take up moisture.[8]

ALFISOLS

Although the Alfisols and Ultisols both have an argillic horizon, they differ appreciably in base status. Ultisols tend to be infertile and contain few weatherable minerals capable of releasing base nutrients, whereas Alfisols contain a moderate to high reserve of bases. This difference may result from: (1) more youthful unweathered minerals in the solum of the Alfisols, or (2) bases being added to the Alfisols by wind and/or water action. The threshold base saturation minimum of these soils is 35 percent at a depth of 50″ from the surface.[9]

Regardless of geographic location, climatic regime, or vegetative cover, practically all Alfisols also have an ochric epipedon. This surface horizon may exhibit a wide range of characteristics and is found where conditions are not conducive to the formation of other more striking and taxonomically more significant diagnostic epipedons.

A description of an Alfisol profile follows. This particular soil is an Albaqualf, which seasonally has groundwater perched above a slowly permeable argillic horizon. The soil normally exhibits an abrupt textural change between either the ochric epipedon and albic or argillic horizon. This particular pedon is formed from Wisconsin loess in Appanoose County, Iowa, under a former hardwood forest cover.

Note in the profile (1) the increase in clay content from the A2 to the B2tg horizon, (2) the light-colored A horizons, and (3) the presence of clay skins in the B horizons.

The Alfisols are divided into five suborders: Aqualfs, Boralfs, Udalfs, Ustalfs, and Xeralfs (Figure 10.6). The *Aqualfs* (L. *aqua*, water) frequently are associated with a seasonally fluctuating water

[8] National Cooperative Soil Survey, *Soil Taxonomy*, p. 3-8.

[9] Exceptions to the depth value are permitted whenever a lithic or paralithic contact is within 50″ of the surface, or when the mean annual soil temperature is less than 57°F.

Horizon	Depth (Inches)	
Ap	0-7″	Dark gray (10YR 4/1) silt loam, light brownish gray (10YR 6/2) dry; few fine faint olive brown (2.5Y 4/4) mottles; weak very fine platy structure parting to fine granular structure; friable; very strongly acid; abrupt smooth boundary.
A2	7-15″	Light brownish gray (10YR 6/2) silt loam, white (10 YR 8/2) dry; moderate fine platy structure; friable; few fine dark concretions; very strongly acid; clear smooth boundary (about 17 percent clay).
AB	15-17″	Pale brown (10YR 6/3) light silty clay loam, white (10YR 8/2) dry; few fine distinct yellowish brown mottles; weak fine subangular blocky structure; friable; thick continuous silt coats; strongly acid; abrupt smooth boundary.
B21tg	17-25″	Dark grayish brown (10YR 4/2) interiors, dark grayish brown (10YR 4/2) and very dark grayish brown (10YR 3/2) ped faces in the upper part; silty clay, few fine distinct yellowish brown, brown, and yellowish red mottles; strong fine and medium subangular blocky structure; very firm; thick discontinous very dark gray clay skins; common fine dark concretions; very strongly acid; clear smooth boundary (about 50 percent clay).
B22tg	25-40″	Dark grayish brown (2.5Y 5/2) silty clay with some brown in lower part; few fine distinct yellowish brown, strong brown, and yellowish red mottles; weak medium prismatic structure parting to moderate medium subangular blocky structure; firm; thin discontinuous clay skins; few fine dark concretions; medium acid; gradual smooth boundary.
B3tg	40-56″	Grayish brown (2.5Y 5/2) silty clay loam; common fine yellowish brown mottles; very weak medium prismatic structure; firm; common very dark gray stains on peds; thin discontinuous clay skins; few fine hard concretions; medium to slightly acid.

table and show signs of wetness, such as mottles and iron manganese concretions larger than 2mm (.08″). *Boralfs* (Gr. *boreas*, northern) are found primarily in cool or cold climates, and have albic horizons. *Udalfs* (L. *udus*, humid) occur in humid climates with a short moisture-deficient period. They tend to be brownish or reddish in color and are relatively freely drained. *Ustalfs* (L.*ustus*, burnt) are mostly reddish-colored Alfisols of warm subhumid to semiarid regions. They have an ustic moisture regime (a warm rainy season), but in most of them moisture moves through the soil to deeper layers only in occasional years. Some have a calcic horizon below or in the argillic horizon, if carbonates are in the parent material or in the dust that settles on the soil. The *Xeralfs* (Gr. *xeros*, dry) are mainly associated with a Mediterranean climate. They become dry for extended periods during the summer, but in an occasional year and in some cases every year moisture moves through the soil in winter to deeper layers.

ULTISOLS

In contrast to the Alfisols, the Ultisols generally are of low fertility due to a small supply of bases. These are the most weathered of all the midlatitude soils and are found on surfaces that are dominantly Pleistocene or older. Glacial ice did not rest on these soils during the Pleistocene period. Consequently, they have been subject to weathering processes for a much greater length of time than have the soils of humid regimes on their northern boundaries. Extensive chemical weathering of their solum has led to a removal of bases; yet contrary to expected increased nutrient availability with depth, the reverse is true:

> The roots of trees go several meters deep in many soils, and the bases they extract at these depths are eventually returned to the surface of the soil. Before the bases can be moved very deeply into the soil, they are again taken up by the roots. Thus, the bases are held against leaching ordinarily by the plants. The supply is partly a function of the species of plants. Some collect large amounts of bases, and the soil below one of these may be well supplied during the mature life of the plant. But the maintainance of bases in the surface horizon is at the expense of the supply in the deeper horizons.[10]

Once the native plant cover is removed, as for planting crops, the meager store of nutrients is rapidly lost and potential crop yields decrease dramatically. Only through conscientious fertilization programs can permanent agriculture be practiced on such soils.

Two features found in some Ultisols are plinthite and fragipans.[11] *Plinthite* (Gr. *plinthos*, brick) is related to warm, humid climates and is most extensively developed in tropical soils. It is an iron-rich mixture of clay with quartz that commonly occurs as dark red or brown mottles and will change irreversibly to ironstone hardpans or irregular aggregates upon repeated wetting and drying. Fragipans impede the downward percolation of water and extension

[10]Smith, "Soil Classification," pp. 76-77.
[11]See Chapter 9 for explanation of fragipan.

ALFISOLS

Mineral soils that have either an argillic or natric horizon, and either a frigid temperature regime or base saturation 50" below the argillic horizon of 35% or more.

AQUALFS

Alfisols that have characteristics associated with wetness.

Albaqualfs have: (1) an abrupt textural change between an ochric epipedon or an albic horizon and an argillic horizon, and (2) low permeability in the argillic horizon.

Duraqualfs have a duripan.

Fragiaqualfs have a fragipan.

Glossaqualfs have an albic horizon.

Natraqualfs have a natric horizon.

Ochraqualfs have an ochric epipedon.

Plinthaqualfs have plinthite as a continuous phase or constituting > 50% of the matrix within some subhorizon.

Tropaqualfs have a mean annual soil temperature ⩾8°C and seasonal soil temperature range at 50cm that is less than 5°C.

Umbraqualfs have an umbric epipedon.

BORALFS

Freely drained Alfisols of cool places.

Cryoboralfs have a cryic temperature regime and an argillic horizon.

Eutroboralfs have a base saturation > 60% in all subhorizons of the argillic horizon, and are dry in some horizon at some time in most years.

Fragiboralfs have a fragipan.

Glossoboralfs are either never dry or have base saturation < 60% in some subhorizon of the argillic horizon.

Natriboralfs have a natric horizon.

Paleboralfs have an argillic horizon deeper than 60" below the mineral surface.

UDALFS

These are brownish to reddish, freely drained Alfisols.

Agrudalfs have an agric horizon.

Ferrudalfs have a discontinuous albic horizon and have iron enriched nodules in the argillic horizon.

Fragiudalfs are other udalfs that have a fragipan.

Fragiglossudalfs have tongues of albic materials in the argillic horizon and have a fragipan.

Glossudalfs have tongues of albic materials in the argillic horizon and no fragipan.

Hapludalfs are other Udalfs.

Natrudalfs have a natric horizon.

Paleudalfs have an argillic horizon with a clay distribution such that the percentage of clay does not decrease by as much as 20% of the maximum within a depth of 60" from the soil surface.

USTALFS

Have a ustic moisture regime, an epipedon that is both massive and hard when dry, or a calcic horizon within 60" of the surface.

Durustalfs have a duripan within 40" of the surface.

Haplustalfs are relatively thin, reddish to brownish in color that have no petrocalcic horizon within 60" of the surface.

Natrustalfs have a natric horizon.

Paleustalfs have a petrocalcic horizon within 60" of the surface or a dense argillic horizon.

Plinthustalfs have plinthite forming a continuous phase.

Rhodustalfs have argillic horizons redder than 5YR.

XERALFS

Alfisols that have a xeric moisture regime, or an epipedon that is both massive and hard when dry.

Durixeralfs have a duripan within 40" of the surface.

Haploxeralfs are relatively thin reddish to brownish Xeralfs.

Natrixeralfs have a natric horizon.

Palexeralfs have a petrocalcic horizon within 60" of the surface or a solum thicker than 60".

Plinthoxeralfs have plinthite that forms a continuous phase.

Rhodoxeralfs have argillic horizons with colors redder than 5YR.

Figure 10.6 Suborders and Great Groups of the soil order Alfisol.

of plant roots. Water either stands above these pans in soils of level areas or moves laterally along the top of the pan if the soils are sloping.

The Ultisols contain five suborders: Aquults, Humults, Udults, Ustults, and Xerults, (Figure 10.7). The *Aquults* (L. *aqua*, water) are wet Ultisols that experience a fluctuating water table and are usually gray or olive in color. *Humults* (L.*humus*, earth) are more or less freely drained, humus-rich Ultisols. *Udults* (L.*udus*, humid) are also relatively freely drained, but are humus poor. They generally have light-colored (grayish) epipedons resting on a yellowish brown to reddish argillic horizon. The *Ustults* (L.*ustus*, burnt) are relatively freely drained Ultisols of warm regions with high rainfall but a pronounced dry season. They have little organic carbon and most have reddish colors. *Xerults* (Gr. *xeros*, dry) are found in Mediterranean climates. They contain moderate or small amounts of organic matter, and have ochric epipedons resting on a brownish to reddish argillic horizon.

LAND UTILIZATION AND MANAGEMENT PROBLEMS

It is difficult to generalize about the land-use patterns and management problems that occur in conjunction with Alfisols and Ultisols. These soils account for approximately 23 percent of the world total,[12] exhibit a wide range in base status, and have considerable variation in length of growing season. Hence, there are many forms of land use on Alfisols and Ultisols when considered in combination at the order level. The basic concept of these soils, however, relates to their development in humid, middle latitude locations. Although they are found in both the tropics and seasonally arid subtropical climates, their utilization and limitations will be considered here only in the areas that are characteristic of the basic type.

The Alfisols of the middle latitudes are noted for a relatively high degree of base saturation and fertility. They support some of the earth's most intensive forms of agriculture. In the United States, the well-known agricultural region called the "corn belt" occurs mainly on Alfisols and Mollisols. Over the years this area has yielded as much as two-thirds of the nation's corn, oats, and soybeans, and nearly one-half of its alfalfa.[13] This region provides a wide choice of

[12]Alfisols make up 14.7 percent and Ultisols, 8.5 percent.

[13]W. H. Pierre and F. F. Riecken, "The Midland Feed Region," *The 1957 Yearbook of Agriculture: Soil* (Washington, D.C.: U.S. Govt. Ptg. Office, 1957), p. 535.

Figure 10.7 Suborders and Great Groups of the soil order Ultisol.

ULTISOLS

Mineral soils of the mid to low latitudes that have a horizon in which there are translocated silicate clays but with only a small supply of bases.

AQUULTS

These are Ultisols of wet places, when the ground water is very close to the surface part of the year.

Albaquults have a marked increase in the percentage of clay in the upper part of the argillic horizon.

Fragiaquults have a fragipan.

Ochraquults have a ochric epipedon.

Paleaquults are found on old land surfaces and have thick, mottled argillic horizons.

Plinthaquults have plinthite present.

Tropaquults are found in intertropical regions.

Umbraquults are dark colored.

HUMULTS

These are relatively freely drained, humus rich Ultisols of mid or low latitudes.

Haplohumults have some weatherable minerals in the argillic horizon.

Palehumults are reddish Humults of old stable surfaces, with thick argillic horizons and few weatherable minerals.

Plinthohumults have plinthite.

Sombrihumults have a sombric horizon.

Tropohumults are found in intertropical regions and have thin argillic horizons.

UDULTS

These are relatively freely drained, humus-poor Ultisols of humid climates.

Fragiudults have a fragipan.

Hapludults have an ochric epipedon and a relatively thin argillic horizon.

Paleudults are relatively freely drained Udults on very old stable land surfaces.

Plinthudults contain plinthite.

Rhodudults are freely drained Udults with dark colors throughout.

Tropudults are relatively freely drained Udults and found in intertropical regions.

USTULTS

Ultisols of warm regions with high rainfall but with a pronounced dry season.

Haplustults have a relatively thin argillic horizon and 10% or more weatherable minerals.

Paleustults have thick argillic horizons and few weatherable minerals.

Plinthustults have plinthite present.

Rhodustults have dark or dusky red argillic horizons and dark colored epipedons.

XERULTS

These are relatively freely drained Ultisols of Mediterranean climates.

Haploxerults have thin or moderately thick argillic horizons and/or appreciable amount of weatherable minerals.

Palexerults have thick argillic horizons and few weatherable minerals.

soil and land-management systems, including cash grain, dairying, and livestock feeding operations. Regardless of the system used, lime, legumes, and phosphates give the maximum plant yield.

Because of the intensive grain cropping, however, many soils have lost one-fourth to more than one-third of their original content of nitrogen and organic matter. With less than 10 percent of the harvested cropland in legume hay and approximately 50 percent in corn, the legumes cannot supply the nitrogen needs of the grain crops in the rotation.[14]

The decrease in available nitrogen in these soils necessitates heavy applications of commercial nitrogen fertilizer. Experiments have shown that 80 to 100 pounds of fertilizer nitrogen per acre can, on certain Alfisols, double corn yield.

To maintain an economical operation on Alfisols, the farm manager not only must be concerned with the nitrogen content of his soils, but must also consider: (1) the addition of frequently deficient phosphorus and potassium, and (2) economizing on tillage operations. Even with increased applications of phosphorous and potassium, the heavy demand for these nutrients and their loss through leaching still pose a problem to farmers in attempting to achieve potential crop yields. Tillage operations present an entirely different set of problems. Excessive tillage results in a decline of soil organic matter, a deterioration of tilth, and an increase in soil erosion. To counteract these problems, some farm managers have:

. . . adopted new methods, which are aimed at minimum tillage for corn in order to save time and money and to maintain better tilth, reduce soil compaction, and enhance water absorption. Among them are the substitution of chemicals for at least one cultivation in the control of weeds; the use of subsurface tillage instead of plowing, with residues left on the surface; and the so-called plow-plant method, in which the corn is planted in the wheel tracks immediately after plowing and the seedbed between the rows is kept loose with minimum tillage.[15]

These methods, combined with contour tillage, are especially important on sloping soils that are subject to erosion. A combination of subsurface tillage with a mulch of crop residue can, according to estimates, reduce the erosion loss that accompanies traditional plowing and tillage by a minimum of one-half.

[14]Ibid., p. 58.
[15]Ibid., p. 539.

Compared with the more northern Alfisols, the midlatitude Ultisols have deeper and more thoroughly weathered solums with much lower base saturation, are acid in reaction, and are relatively infertile. They give a distinct advantage to the farm manager, however, in that they experience a lengthy growing season (frost-free days range from 200 to 260 in the United States) normally accompanied by abundant rainfall.

Land-use forms on the Ultisols are many. The mild climate is favorable to the cultivation of crops such as cotton and peanuts that cannot be grown farther north. Under proper fertilization programs these soils also may have productive yields of corn, oats, tobacco, and forage crops. Combined with mild winters, forage crops permit the grazing of livestock throughout the year. In addition, the native forest cover represents a huge potential for the pulp and plywood industries, and extensive tracts of land have been set aside for tree plantations.

The same climate characteristics that make the Ultisol region attractive for agriculture have been largely responsible for the soil's low fertility and mineral deficiencies. Excessive leaching of the profile has led to a removal of minerals from the solum and a displacement of the bases by hydrogen ions, resulting in a pH that in the United States averages between 5.0 to 5.5. Along with large moisture surpluses is a relatively high temperature regime that prevents the accumulation of significant amounts of organic matter and nitrogen.

Nitrogen is critical to the production of healthy crops and must be considered in most rotations. Some land managers have planted winter legumes as a cover crop in an attempt to improve soil structure and add nitrogen. However, most farmers prefer to meet the nitrogen deficiency by applying commercial fertilizer. Yield increases for corn, assuming other limiting factors are not present are shown in Figure 10.8. Other crops have responded in a similar fashion. Pearson and Ensminger, in a report on the effect of nitrogen applications on forage crops and the resultant weight gain of grazing cattle, state that each pound of applied nitrogen produced an average 2.7 pounds of beef.[16]

A well-planned fertilization program not only must provide nitrogen but also must reduce acidity and add deficient nutrients. A normal practice on most Ultisols is to provide soil calcium and raise

[16]R. W. Pearson and L. E. Ensminger, "Southeastern Uplands," *The 1957 Yearbook of Agriculture: Soil* (Washington, D.C.: U.S. Govt. Ptg. Office, 1957), p. 585.

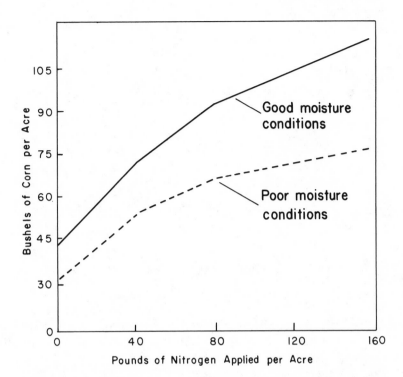

Figure 10.8 Corn yields with nitrogen applications under varied moisture conditions.

the pH level through liming and to supply phosphorus and potassium for most crops. In certain areas where boron, zinc, and sulfur are deficient, their application can increase crop yields significantly.

Oxisols

11

Oxisols are mineral soils that have either an oxic horizon within 80" (2 meters) of the surface or plinthite that forms a continuous phase within 12" (30cm) of the mineral surface of the soil. *Oxic horizons* are subsurface horizons at least 12" thick consisting of a mixture of hydrated oxides of iron and/or aluminum, often in an amorphous state, and containing variable amounts of 1:1 lattice clays. Few weatherable minerals remain in the oxic horizon, except for such highly insoluble materials as quartz sand. Hence, there is little additional release of bases by further mineral alteration, and the cation-exchange capacity tends to be low.

Plinthite may occur with an oxic horizon or by itself. This sesquioxide-rich,[1] humus-poor mixture of clay with quartz changes, as we know, to ironstone hardpans or irregular aggregates upon repeated wetting and drying. (See Ultisols in Chapter 10 for full explanation of plinthite.)

The Oxisols have experienced the greatest degree of mineral alteration and profile development of any soil. They exist primarily on ancient landscapes in the humid tropics. Seldom found over broad contiguous areas, they are likely to occur with more youthful Ultisols, Entisols, and Vertisols. There are two main regions of Oxisol concentration: in South America surrounding the alluvial soils of the Amazon and in equatorial Africa (Figure 11.1). Less significant areas are found in parts of India, Burma, and Southeast Asia.

[1] Remember that a *sesquioxide* is an oxide with one and a half oxygen atoms to every metallic atom.

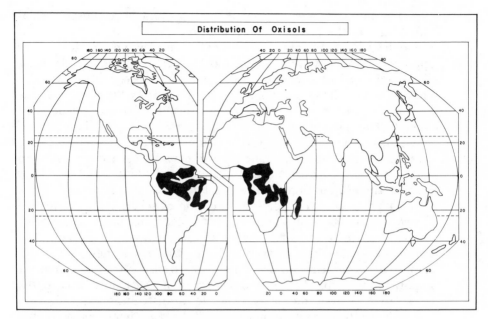

Figure 11.1 World distribution of Oxisols.

CLIMATE AND NATIVE VEGETATION

The climate and vegetation of the intertropical areas exert a profound, interacting influence upon the region's pedogenic activity. No other portion of the globe has experienced such a long period of uninterrupted weathering processes and vegetative evolution. Unlike the middle latitudes where comparatively recent glaciation greatly modified large portions of the surface and restricted the development of certain plant species, the more stable climate of the tropics has encouraged the growth of the world's most heterogeneous plant formations, as well as the deep and thorough alteration of the mineral crust.

Oxisols occur in climates with these unique characteristics: (1) intense solar radiation, (2) a relatively uniform diurnal (daily) photoperiod (lengths of alternating periods of light and dark), (3) isothermal to near-isothermal monthly temperatures, (4) high potential evapotranspiration rates, and (5) varied amounts of precipitation, including seasonal aridity. The extensive range of moisture conditions under which oxic horizons exist has led numerous researchers to conclude that present precipitation patterns are not necessarily responsible for their distribution—many oxic horizons are found in areas of limited rainfall. Rather, it is believed that the most extensive areal development of the oxic horizon may have taken place under paleo-climates of much higher rainfall.

The humid tropics are noted for large annual surpluses of solar radiation. On the average the equator receives 2.5 times as much as the poles.[2] Yet, seldom do temperature maximums in the tropics approach those of the midlatitudes. This apparent anomaly is explained by atmospheric circulation and ocean currents. Atmospheric circulation carries away approximately 80 percent of the tropics' surplus energy and prevents its accumulation at the point of receipt. The energy that is transported poleward is in the forms of *latent heat* (utilized in the vaporization of water) and *sensible heat* (absorbed in air masses). Latent heat may be carried poleward to make up for inequal heating at different latitudes, but this energy used in the evaporation process has two additional functions in the hot, humid tropics: (1) since this portion of solar energy produces a change in the state of water—for example, from liquid to vapor—it does not increase air temperature at the earth's surface; and (2) in converting water into a gaseous form, the atmosphere is given a supply of potential rainfall. Therefore, transporting latent heat is important not only to the temperatures of the tropics but to the moisture it receives and makes available for distribution as well.

Due to the energy exchanges that take place under a remarkably uniform photo-period, humid equatorial locations experience little day-to-day variation in temperature (Figure 11.2). This uniformity induces a monthly moisture demand (PE) that is also highly uniform. The water budget graph in Figure 11.2 for Uapes, Brazil, shows this station's monthly PE does not vary from the mean of 4.71" by more than 1/2" in any month.

The total "annual" rainfall in the tropics does vary, however, from place to place, and fluctuates considerably at the same station as well. When the maximum annual rainfall is expressed as a percentage of the mean, the minimum averages about 60 percent and the maximum around 150 percent—for example, if the mean annual rainfall is 100" during some years precipitation may be as low as 60" while in others it may be as high as 150.[3] As far as soil formation is concerned, the way in which rainfall arrives is as important as the total annual amount. The higher the rainfall per day, the less valuable it becomes to the soil. Within the tropics a daily fall as high as 6" to 8" is expected at least once every two years.[4] In Indonesia a daily

[2] R. G. Barry and R. J. Chorley, *Atmosphere, Weather, and Climate* (New York: Holt, Rinehart, and Winston, Inc., 1970), p. 33.

[3] E. C. Jo Mohrand and F. A. Van Baren, *Tropical Soils* (London: Interscience Publishers, Ltd., 1954), p. 31.

[4] J. Chang, "The Agricultural Potential of the Humid Tropics," *The Geographical Review*, 58 (1968), 358.

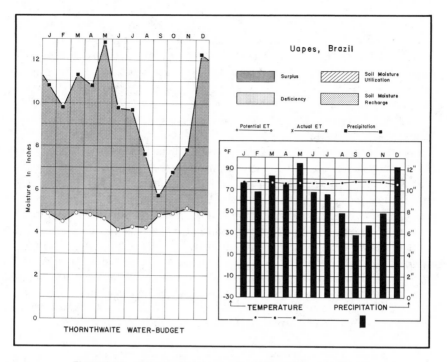

Figure 11.2 Water budget: Uapes, Brazil (based on normal data).

maximum of 27.6″ was recorded at Ambon.[5] Obviously, the soil cannot hold that much water, and a good deal of the surplus becomes overland flow with high erosive power—especially on cultivated plots.

In 1951 measurements taken in soils that were cultivated without precautions against erosion in the Casamance River Basin in West Africa showed an annual soil loss of 1,600 tons/mi[2] on a slope of 1 percent, and 2,500 tons/mi[2] on a slope of 1.5 percent, under peanuts, and of 1,400 tons and 1,960 tons/mi[2] respectively under upland rice. Although Casamance lies outside of the tropical rain forest climate, the results indicate the phenomenal erosive power that is available within the humid tropics.[6]

In tropical regions where rainfall is heavy and reliable throughout most of the year, a native vegetation called the *tropical rain forest* occurs. Each of the three distinct types—American, African, and Indo-Malaysian—has its own plant families and genera, yet all

[5] Mohrand and Van Baren, *Tropical Soils,* p. 38.

[6] *The Geographical Review,* 58, 358.

have evolved plants with similar structures and appearances. The tropical rain forest vegetation is distinctly different, however, from middle latitude forests in both morphology and number of species.

The tropical rain forest is considered the most heterogeneous of all living assemblages in the world, with as many as 3,000 different species of trees existing within a one-square-mile area. The forest also contains the earth's most diverse groups of epiphytes[7] and parasites.[8] The woody plants of this vegetative community are primarily evergreens in that they lose their leaves and grow new ones simultaneously. The forest presents an array of stratified canopies, for example, species tend to crown at various heights depending upon sensitivity to light (Figure 11.3). Those demanding the most sunlight are normally the tallest and characteristically have broad umbrella-shaped crowns. Lower strata species tend to be less tolerant to light and thrive under the shade of their taller neighbors. They are spaced closer together and develop conical-shaped crowns. These are mostly thin-barked trees with long straight trunks, few branches beneath the crown, and buttressed bases (for the larger members).

The very complex canopy structure with crowns occurring at distinct levels prevents most direct solar radiation from reaching the

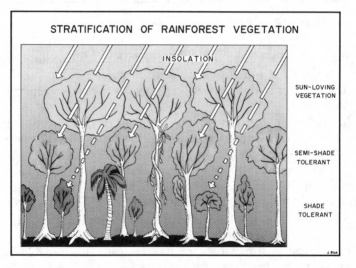

Figure 11.3 Structure of a mature tropical rain forest.

[7] Epiphyte—a plant that grows on another plant but is not parasitic, obtaining its essential foods from the atmosphere.

[8] Parasite—a plant that grows on another plant and lives at the expense of host plant.

surface soil of the forest floor. Only an estimated one percent of the energy, reaching the uppermost canopy of a mature stand, is capable of penetrating through the foliage to the soil surface. With such small light intensity, most low-growing plants cannot survive. Therefore, the undergrowth of the rain forest, except for mosses and ferns, is of little significance.

Vegetation of the tropical rain forest has several important functions related to pedogenesis and soil utilization. It serves as a: (1) nutrient reservoir, (2) humus source and food for soil inhabitants (through leaf fall and the like), and (3) protective soil cover. Under conditions of high annual precipitation, the continued leaching of the soil permits the rapid removal of the soluble base nutrients. The primary way in which bases are conserved in such climates is through temporary immobilization. This is largely accomplished in the rain forest by vegetation, which in the act of withdrawing life-support elements from the soil incorporates nutrients within their cellular structure. Thus, the standing biomass contains a long-term accumulation of material that otherwise might have been lost from the system. Burning this vegetation provides a source of fertilizer as organo-mineral complexes are oxidized. Frequently, this is the only form of soil enrichment available for many isolated fields.

Humus would not exist without an organic source to provide the raw substance from which it originates. The tropical rain forest is well designed to supply a continual and bountiful amount of such material. Rather than a complete seasonal shedding of leaves, the evergreens tend to maintain a continuous pattern of shedding older, inefficiently functioning leaves and at the same time replacing them with younger, more vigorous members. Thus, as leaves fall to the ground and aging plants die, the forest floor is covered with organic litter that, depending upon moisture conditions, is broken down rapidly by microorganisms and insects into various states of decay. The oxidation of organic matter under conditions of good soil drainage is often so rapid, however, that little is left to be incorporated in the soil.

The natural vegetation of the rain forest also protects the soil from accelerated erosion and surface dryness. Unprotected surfaces are affected by the high energy impact from raindrops and considerable overland flow, which detaches soil particles and washes them downslope at a disproportionately high rate. With a forest cover, raindrop impact not only is reduced, but the velocity of the surface flow is slowed as well. The canopy also prevents the intense heating of the soil surface, thus reducing the soil moisture demand and the possibility of the dehydration of epipedon constituents.

PEDOGENESIS

Geomorphic evidence suggests that oxic horizons tend to be associated with surfaces that are mid-Pleistocene in age or older, occur at low elevations (primarily below 6,000'), and generally have parent ferromagnesian (basic) rock material. Most of these soils exist on surfaces exhibiting minimal relief—they are normally level or have relatively gentle slopes. Their position is one in which weathered sediments could have been deposited, but recent unweathered sediments could not accumulate on them, and groundwater would not move laterally to them from an area where fresh rock is weathering.[9]

The processes leading to the formation of oxic horizons are called *laterization*. In these weathering processes silica is removed from the primary silicates and part of the quartz (when present), and alkali and alkaline earth metals also are removed. The soluble soil constituents are almost completely washed out by leaching, while the silica can combine with alumina to produce clay minerals, kaolinite being the final product. The remaining silicate is leached out of the profile. This process is called desilication, and consists of loss of silica and a simultaneous gain of weathering products of increased stability, among which iron oxides and hydroxides (and sometimes quartz) are important.[10]

Certain Oxisols that have, or appear to have had, a fluctuating groundwater table contain plinthite (see Chapter 10 under Ultisols). Schuylenborgh recognizes three distinct stages in plinthite development.[11] Stage one is associated with free or nearly free drainage. The atmospheric effects of heating and cooling, and wetting and drying, combined with hydration and hydrolysis, carbonation, oxidation and dissolution, and the influence of soil fauna and flora, all unite to disintegrate and decompose the parent rock material. During this period secondary minerals (for example, kaolinite, gibbsite, geothite) are formed, and bases, silica, and part of the aluminum are removed through active leaching processes. As weathering continues, the amount of clay minerals and amorphous sesquioxide colloids increases, and the erosion base level is approached—leading to the second stage.

---

[9] National Cooperative Soil Survey, *Soil Taxonomy* (Washington, D.C.: U.S. Govt. Printing Office, 1970), pp. 3-28 and 3-29.

[10] J. Van Schuylenborgh "Investigations on the Classification and Genesis of Soils Derived From Andesitic Tuffs Under Humid Tropical Conditions," *Netherlands Journal of Agricultural Science*, 5, (1957), 195-210.

[11] Ibid.

In the second stage the amount of rainwater permeating the soil lessens sufficiently to prevent free drainage of all the water during the wet season. This results in a seasonal groundwater table within the profile, even though it may be removed during the dry season. As the soil undergoes the repeated appearance and disappearance of groundwater, weathering and soil-forming reactions are characterized by alternating reductive and oxidative conditions. The soil material in the fluctuating water table will begin to show mottling associated with the segregation of iron, but the epipedon may still retain a uniform color.

After continued weathering the third stage of plinthite formation begins as drainage and evapotranspiration become incapable of removing all of the subsurface water—even in the dry season—and a permanent groundwater table is established. Under water-saturated conditions and low oxygen, normal gley formation occurs within a zone of constant reduction. The primary result is the reduction of iron to its mobile ferrous form. A fluctuating groundwater table still exists, even though the highest water level normally is lowered upon weathering. This eventually results in a thickening of the iron enriched mottled horizon that is called plinthite. Portions of the original mottled clay now are solely moistened by capillary rise from the water table, and the upper parts actually may become dry. Subsequent drying of the mottled zone can result in a cementing of the matrix materials through the dehydration and crystallization of amorphous iron and iron oxyhydrates into cryptocrystalline and crystalline hydroxides and oxides—thus creating an ironstone hardpan.[12]

The following is a description of an Oxisol profile in Mysore State, India:

Horizon	Depth (Inches)	
A	0-18 "	Reddish brown (5 YR 4/3), silty clay, strong, fine, granular structure, friable, iron concretions, smooth boundary.
B	18-48"	Reddish brown (5 YR 4/3), clay loam, strong, medium, subangular blocky structure, firm, nonplastic and nonsticky, bigger iron concretions, diffuse boundary.
C	48"+	Reddish brown (5 YR 4/3), clay loam, reticulately mottled, strong, medium, subangular blocky structure, firm, slightly sticky, rounded red and black iron concretions, plinthite.

[12]*Ironstone* refers to plinthite that has irreversibly hardened.

"B" has all of the characteristics of an oxic horizon, containing a mixture of hydrated oxides of iron and aluminum, 1:1 lattice clays, and insoluble quartz sand. Cation-exchange capacity is low in this soil, and plinthite is present. The plinthite is normally soft when not exposed, but, as we have described, changes irreversibly to ironstone hardpans or irregular aggregates with repeated wetting and drying. In this area dried bricks made from such soil are used for building materials.[13]

Subdivision of the Oxisols rests primarily on characteristics that indicate annual and seasonal moisture. Moisture is an important variable in tropical soils. It may determine the amount of organic matter, nutrient availability, and the degree of plinthite development, and it is an overall index of current chemical weathering.

There is a direct relationship between soil moisture status and organic matter—the latter increases with abundant water and deficient oxygen when microbial activity is limited. Base saturation, on the other hand, bears an inverse relationship to soil moisture. Under continued leaching, soluble nutrients are removed from the profile through subsurface drainage. As we know, plinthite is related to impeded drainage and a fluctuating groundwater table. All of these activities, except for organic matter accumulation, involve pedochemical reactions whose intensity is directly related to the soil moisture regime.

The Oxisols are separated into five suborders: Aquox, Humox, Orthox, Torrox, and Ustox (Figure 11.4). The *Aquox* (L.*aqua*, water) are wet Oxisols that have plinthite within 12″ of the surface and are saturated with water at this depth during part of the year. *Humox* (L.*humus*, earth) are normally found in high altitudes and have relatively high contents of organic matter. Yellowish to reddish in color, the *Orthox* (Gr. *orthos*, true) have short or no dry seasons. The *Torrox* (L.*torridus*, hot and dry) are the Oxisols of arid climatic regimes. *Ustox* (L.*ustus*, burnt) are red in color and dry for extended periods, although they are moist for at least 90 days per year.

LAND UTILIZATION AND MANAGEMENT PROBLEMS

In general, the soils of the tropics tend to be of low fertility, with the exception of some that are alluvial or volcanic in origin. Their properties are unique from those of the middle latitudes and

[13]Seshagiri Rao, "Pedogenesis of Some Major Soil Groups in Mysore State, India," *Soils and Tropical Weathering: Proceedings of the Bandung Symposium* (Paris: UNESCO, 1971), pp. 79-84.

OXISOLS

Mineral soils that have an oxic horizon within 80″ of the surface or plinthite that forms a continuous phase within 12″ of the mineral surface of the soil.

AQUOX

Oxisols that have plinthite forming a continuous phase within 12″ of the surface; or have an oxic horizon that has characteristics associated with wetness.

Gibbsiaquox have no plinthite forming a continuous phase within 12″ of the surface and have cemented sheets containing ≥ 30% gibbsite.

Ochraquox have no plinthite forming a continuous phase within the upper 40″ and have an ochric epipedon.

Plinthaquox have an ochric epipedon and plinthite that forms a continuous phase within the soil between 12″ to 50″.

Umbraquox have an umbric or histic epipedon and neither a plinthite formation within 50″ of the surface or gibbsite.

HUMOX

Oxisols that are always moist or have no period when the soil is dry below the surface 7″ for 60 days or more. They have high contents of organic matter and a low supply of bases.

Acrohumox are the most intensely weathered Humox. They have lost virtually all ability to retain bases in their mineral fraction.

Gibbsihumox have within 40″ of the surface, cemented sheets or a subhorizon that contains gibbsite.

Haplohumox have an oxic horizon and a small cation-exchange capacity in their mineral fraction.

Sombrihumox have an oxic horizon with a subhorizon that is darker in color and contains more organic carbon than the overlying horizon.

ORTHOX

Oxisols exclusive of Aquox that have short or no dry seasons. They are most common near the equator.

Acrorthox have lost virtually all ability to retain bases in their mineral fraction.

Eutrorthox are Orthox with relatively high base status.

Gibbsiorthox have gibbsite within 50″ of the surface.

Haplorthox have few bases, but more than the Acrorthox, and the clays have a modest cation-exchange capacity.

Sombriorthox have an oxic horizon with a subhorizon that is darker in color and contains more organic carbon than the overlying horizon.

Umbriorthox are Orthox that have an umbric epipedon or appreciable amounts of organic matter.

TORROX

These soils are usually dry in most years in all parts of the soil. These are the Oxisols of arid climates. (No subdivision of Torrox is presently recognized.)

USTOX

Oxisols that have some subhorizon below the surface 7″ that is dry for at least 60 consecutive days annually. They tend to be found near the Tropics of Cancer and Capricorn.

Acrustox are red or dark red Ustox that have virtually no ability to retain bases in their mineral fraction.

Eutrustox have a mollic or umbric epipedon, an appreciable supply of bases, and moderate to high base saturation in the oxic horizon.

Haplustox are red or dark red Ustox that have clays with modest cation exchange capacity, but have low base saturation in some or all parts of the oxic horizon.

Sombriustox have an oxic horizon with a subhorizon that is darker in color and contains more organic carbon than the overlying horizon.

Figure 11.4 Suborders and Great Groups of the soil order Oxisol.

require special management techniques to realize their yield potential. Under a rain forest cover, the fertility of humid tropical soils is maintained in a delicate equilibrium. Organic matter falling from the trees is decomposed, releasing nutrients and making them available for plant utilization in a continuous cycle that operates with the same small capital of nutrients. Once the rain forest is cleared by man, this balance is destroyed and soil fertility declines rapidly as oxidation and leaching occurs under high temperatures and rainfall without a protective mantle of natural vegetation.

Considering the limited fertility of most tropical soils, they support a surprising number of varied economic functions. Land use ranges from shifting cultivation on a subsistence level to permanent plantations, and from grazing to lumbering operations. By far the most universal use of land throughout most of the humid tropics, however, has been by the shifting agriculturists (Figure 11.5).[14]

In an area of shifting cultivation, the landscape consists of extensive areas of forest broken by cleared patches where cultivation takes place. There is constant clearing of new fields and abandoning

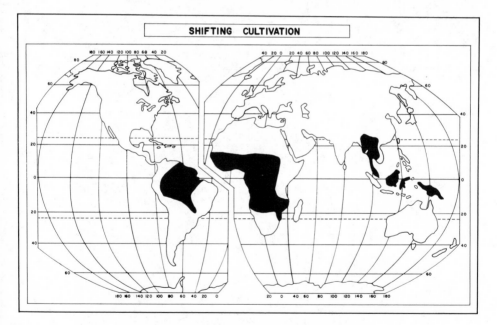

Figure 11.5   World distribution of shifting cultivators.

[14]Shifting agriculture refers to subsistence economic levels and involves the use of fire to clear lands for temporary utilization, for example, 2 to 3 years. It is known by such names as milpa, swidden, ladang, caingin, roza, and others.

of old ones after two or three years of producing vegetables and grain crops. Figure 11.6 illustrates such a unit in the Republic of Panama. The farm covers approximately 250 acres, providing sufficient space for the regeneration of bush fallow and the clearing of new land.

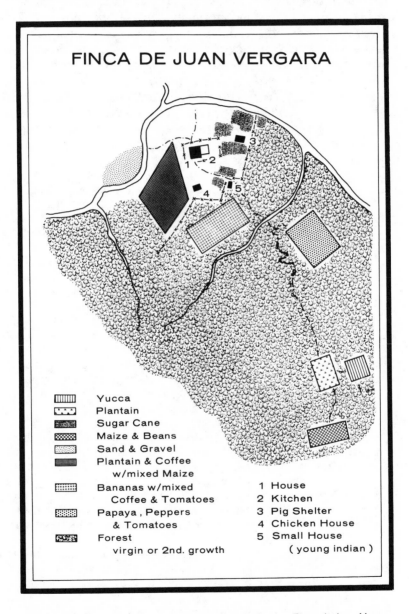

FINCA DE JUAN VERGARA

Yucca
Plantain
Sugar Cane
Maize & Beans
Sand & Gravel
Plantain & Coffee
  w/mixed Maize
Bananas w/mixed
  Coffee & Tomatoes
Papaya , Peppers
  & Tomatoes
Forest
  virgin or 2nd. growth

1 House
2 Kitchen
3 Pig Shelter
4 Chicken House
5 Small House
  (young indian)

Figure 11.6   The field patterns of a shifting agriculturist: Finca de Juan Vergara.

In preparing the land for cultivation, the farmer usually chooses select spots of virgin forest if available. They are easier to clear since the undergrowth is not so dense and tangled. In addition, a forest of tall trees indicates an area that has not recently been used for cultivating crops, hence containing a greater reservoir of stored nutrients. Clearing the land normally involves hacking out lianas, saplings, low underbrush, and either girdling or felling trees. This generally takes place during the period of minimum rainfall, thus allowing the vegetative debris to dry and become more flammable.

Burning is an essential feature of the shifting agricultural system, for in this way only the ground can be cleared of the tremendous mass of felled vegetation. All of the plants in the herb layer are destroyed in the burn, leaving the ground clean. Greenland and Nye give the following comments regarding the effect of burning in these areas:

> All the nutrient elements in the fallow except nitrogen and sulphur are preserved and added to the soil in the ash. The loss of organic carbon and nitrogen that is entailed is often deplored but it is unavoidable. There is no practicable way of incorporating this material in the soil to form stable humus. In grassland, burning is equally essential: a farmer's primary object is to prevent regrowth of the grass by the simplest means, and it is quite impracticable to bury grassland vegetation with a hand hoe.
>
> . . . .
>
> Large quantities of nutrient ions from the standing vegetation and the litter layer are spread in the ash on the surface of the soil in the form of carbonates, phosphates, and silicates of the cations. Nearly all the nitrogen is, however, lost to the atmosphere as ammonia, gaseous nitrogen, or the oxides of nitrogen, and sulphur as sulphur dioxide.
>
> . . . .
>
> . . . plant growth is better in previously heated than in unheated soils. Any seeds present in the soil during the heating are liable to be killed or to have their germination retarded. The supply of available nutrients in the soil solution is increased by heating, and in particular the rate of nitrogen mineralization is generally greater subsequent to the burn than before it. In most instances there is a profound change in the numbers and composition of the microflora of the soil.
>
> The effect of heat on the microbial population is usually referred to as "partial sterilization," and it is similar to effects produced by drying or by treating the soil with antiseptics. The sterilizing treatment leads to an initial decrease in the microbiological population, after which it redevelops, usually with a

modified composition, to a level greater than before. From the soil fertility viewpoint the most important effect of the increase is the change in the rate of nitrogen mineralization. This generally shows parallel changes to the population changes.[15]

After the vegetation has been burned, the soil remains relatively unattended until the approach of the rainy season when crops are planted. The small patches of land upon which crops are grown quite often sustain a variety of intercropped plants, such as maize and beans in the Americas, maize and rice in Africa, and other combinations. In some regions where the rainy season is long enough, two to three crops per year may be harvested.

A second clearing and burning normally takes place after the first year of crop growth. This weakens suckers and destroys seedlings. However, it also encourages the growth of grasses.

There is ample evidence to prove that even when starting with land that either is a virgin or a mature secondary forest yields fall fairly rapidly. "The native farmer, however, does not attempt to crop his land for long enough for declining fertility to be a dominant factor, because of the labour of weeding. It costs him less effort to clear another patch of ground."[16] Encroachment of weeds and grasses is a serious management problem in the tropics. Unlike the grasses of the middle latitudes, which have a high requirement for bases and tend to increase fertility by recycling base nutrients, the grasses of the tropics tend to be much less effective than the forest in preventing nutrient loss through leaching.

As population demands for food increase the pressure on tropical lands, the shifting cultivator has to clear plots in shorter periods of time. He does not appear to respond to such pressure by cropping longer. With less time for the forest to regenerate, the shorter fallow period is conducive to the establishment of savanna. This is accompanied by a decline in the amount of phosphorus and potassium added in the ash, the level of humus, and the amount of nitrogen and phosphorus mineralized from it. The length of time an area can be cultivated without fallowing averages about three to five years. The fallow period normally is approximately three times the length of the cropping period. This is in general agreement with the time estimated to reach a near-equilibrium organic accumulation on forest floors.[17]

[15]D. J. Greenland and P. H. Nye, *The Soil Under Shifting Cultivation* (Bucks, England: Commonwealth Agricultural Bureaux, 1965), p. 67.
[16]Ibid.
[17]*Geographical Review*, p. 357.

Land-use systems in the tropics have often been condemned as exploitive, squandering the natural resources of the area. The main objections concern the loss of forest vegetation, destruction of humus, and loss of nutrients by accelerated leaching. The first objection can be assessed in terms of the demand for agricultural land. Without forest removal, agriculture would be impossible. Second, under an adequate fallow system humus can be maintained at satisfactory levels. Third, although nutrients are definitely leached out of the topsoil during the cropping period, it has not been determined that they are removed beyond the root zone of the fallow. Therefore, the nutrients may still be available for recycling within the biologic system.

Histosols

12

Unlike the other soil orders, the Histosols are not considered primarily mineral, but organic. These soils are commonly called bog, moor, peat, or muck. They are the last in the list of soil orders, last in areal importance, and have been given the least amount of attention in the more recent soil classification systems. In fact, the National Cooperative Soil Survey team considers its current means of differentiating these soils as "provisional."

The Histosols contain organic materials that are either more than 12 to 18 percent organic carbon by weight (depending on the clay content of the mineral fraction and the kind of materials) and well over half organic matter by volume. Unless drained, most Histosols are saturated or nearly saturated with water most of the year. In certain cases they may consist of only an organic mat floating on water. The presence of water is the common denominator in all Histosols regardless of location.

These soils can form in virtually any climate, even in arid regions, as long as water is available. They occur in the tundra and on the equator, and their vegetation consists of a wide variety of water-tolerant plants.

In addition to water, the controlling factors that regulate the accumulation of organic matter include: the temperature regime, the character of the organic debris, the degree of microbial activity, and the length of time in which organic accumulation has taken place. The deposition of organic material within a water-saturated environment rapidly leads to the depletion of oxygen in the wet zone through decomposing aerobic (oxygen-demanding) microorganisms. This eventually results in an anaerobic (without oxygen) milieu, in which the rate of organic matter mineralization is considerably reduced. As a consequence of the impeded decay process, organic materials continue to accumulate. The development of these soils is, therefore, obviously from the mineral surface upward—for example, characterized by a deepening of the organic layer.

The major distinction among the suborders of Histosols relates to the degree of decomposition of organic remains and the soil's moisture. There are four suborders: Fibrists, Folists, Hemists, and Saprists (Figure 12.1).

Fibrists (L.*fibra*, fiber) are Histosols of relatively unaltered plant remains. The plant debris is so little decomposed that it is not destroyed by rubbing, and the botanic origin is clearly evident. They may consist of partially decomposed wood or of the remains of such plants as mosses, grasses, sedges, and papyrus, or of mixtures. The reasons for the preservation of plant remains vary, but the absence of oxygen is probably the most important factor. A lowering of the water table, or its seasonal fluctuation, increases the availability of oxygen, enhances decomposition, and encourages the rapid destruction of fibers. The rate of mineralization varies, however, according to the thermal regime and the nature of the vegetative debris. Bald cypress, for example, has a greater degree of resistance than most woods, grasses, and sedges. As a result, accompanying drainage, mixed deposits will show an increased percentage of volume occupied by Bald cypress, if this vegetative form is present.

Folists are relatively freely drained Histosols consisting of an organic horizon derived from leaf litter, twigs, and branches (but not sphagnum), resting on rock or on fragmental materials made of gravel, stones, and boulders, with the interstices filled or partially filled with organic materials. There is very little evidence of mineral soil development, and plant roots are restricted to organic materials.

Hemists (Gr. *hemi*, half) are Histosols in which the organic material is strongly, but not completely, decomposed. Enough, however, has been broken down to the point that its biologic origin cannot be determined and/or the fibers can be largely destroyed by rubbing between the fingers. These soils are normally found where the groundwater is at or very near the surface most of the year, unless artificially drained. The groundwater levels can fluctuate, but they seldom drop more than an inch below the surface tier. These soils were once classified as bog. They have been found from the equator to the tundra, usually in closed depressions and in broad, very poorly drained flat areas such as the coastal plains.

HISTOSOLS

These are soils that are dominantly organic.

FIBRISTS

These are Histosols that are comprised of organic matter that is decomposed. The botanic origin of the plant remains can be readily determined.

Borofibrists are found in frigid temperature regimes.

Cryofibrists are found in colder regimes than the Borofibrists.

Luvifibrists have *humilluvic* materials.

Medifibrists are found in the middle latitudes.

Sphagnofibrists are mainly derived from various species of Sphagnum and associated herbaceous plants.

Tropofibrists are found in the intertropical areas.

FOLISTS

These are organic soils consisting of leaf litter, twigs, branches, resting on rock or fragmental mineral materials.

Borofolists are found in frigid temperature regimes.

Cryofolists are found in colder regimes than Borofolists.

Tropofolists are found in intertropical areas.

HEMISTS

These are organic soils in which the botanical origin of up to two-thirds of the materials cannot be determined.

Borohemists are found in frigid temperature regimes.

Cryohemists are found in colder regimes than the Borohemists.

Luvihemists have humilluvic materials.

Medihemists are found in the middle latitudes.

Sulfihemists have sulfidic materials within 40" of the surface.

Sulfohemists are acid sulfate soils.

Tropohemists are found in the intertropical regions.

SAPRISTS

The soils consist of almost completely decomposed plant remains.

Borosaprists are found in frigid temperature regimes.

Cryosaprists are found in colder regimes than the Borosaprists.

Medisaprists are found in the middle latitudes.

Troposaprists are found in the intertropical regions.

Figure 12.1 Suborders and Great Groups of the soil order Histosol.

Following is a description of a *Borohemist* (a Hemist of a frigid thermal regime). This soil, in Chase County, Michigan, is in an undrained depression in a sand glacial outwash plain.[1]

Horizon	Depth (Inches)	
Oe1	0-5″	Dark reddish brown (5YR 3/3), dark reddish brown (5YR 3/2) rubbed hemic materials; about 55 percent fiber, about 25 percent rubbed; weak thick platy structure; nonsticky; about 60 percent of the fibers are sphagnum mosses and the remaining portion is herbaceous; extremely acid; abrupt smooth boundary.
Oe2	5-19″	Very dark brown (10YR 2/2) on broken face and rubbed hemic materials; about 40 percent fibers, 15 percent rubbed; weak thick platy structure; nonsticky; 85 percent of fibers are herbaceous, the remaining woody or mosses; extremely acid; clear smooth boundary.
Oe3	19-28″	Dark brown (10YR 3/3) on broken face and rubbed hemic materials; about 50 percent fibers, 25 percent rubbed; weak thick platy structure; nonsticky; 85 percent of fibers are herbaceous with remaining portion woody; few woody fragments 1 to 6 inches in diameter; extremely acid; clear smooth boundary.
Oe4	28-60+″	Dark reddish brown (5 YR 3/3) on broken face and rubbed hemic materials; about 60 percent fibers, 30 percent rubbed; massive; nonsticky; about 90 percent of fibers are herbaceous with a few woody fragments; very strongly acid.

Saprists (Gr. *sapros*, rotten) consist almost entirely of decomposed plant remains. The botanic origin of the materials is, for the most part, obscure and their color is usually black. They occur in areas where the groundwater level fluctuates within the soil and are subject to aerobic decomposition. When Fibrists and Hemists are drained artificially or naturally, they normally will continue to decompose and convert to Saprists.

LAND-USE AND MANAGEMENT PROBLEMS

Where Histosols are drained in such a way as to permit the rapid removal of excess water, yet retain the water level at a relatively

---

[1] O is used to designate an organic horizon; *e* (Oe) represents a Hemic horizon; other organic horizons include *i* (Oi) for fibric, and *a*(Oa) for sapric.

shallow depth, these soils can be utilized very profitably for intensive forms of crop production. The wise management of water is critical. In their original state, these soils are too wet either to support the operation of farm equipment or to produce crops. When they are drained, they become aerobic and may oxidize and subside rapidly. The rate of oxidation and subsidence is very closely linked with the thermal regime in which a particular Histosol occurs. In Florida, subsidence takes place at an estimated rate of $2''$ to $3''$ each decade for Histosols under cultivation. These rates are significantly reduced further poleward where temperatures are lower.

The ease with which plant roots can penetrate, as a result of low bulk density, permits intensive cultivation of a variety of crops on Histosols. The major agricultural limitation in any given area is climate. A late spring or early fall frost in the midlatitudes can create severe economic losses and make it unadvantageous to produce certain crops.

Cabbage, carrots, celery, cranberries, mint, onions, potatoes, and a variety of root crops have proven successful on these soils if managed properly. In addition to sophisticated drainage systems, potential crop production can be realized only when the farm manager follows a conscientious fertilization program and protects his soil from erosion and fire hazards. Organic soils normally require little or no nitrogen fertilization, but they do need relatively heavy applications of phosphorus and potassium. In addition, calcium and magnesium are frequently in short supply and must be compensated by liming. Deficiencies of secondary elements depend largely upon the crop to be cultivated and the character of the organic debris. The more common deficiencies that occur include copper, zinc, and boron (especially for celery and clover).

When the Histosols become dry through excessive drainage or during prolonged drought periods (when the water table drops), they are subject to fire damage. The dried organic material is a virtual tinderbox that can be ignited by careless human acts or by natural occurring phenomena such as lightning. Once a fire is started these soils may smolder for months. The degree of soil destruction from uncontrolled fires can be extensive. In fact, one of the current theories regarding the formation of Drummond Lake in the Dismal Swamp of North Carolina involves the concept of fire "burning out" a depression in relatively thick deposits of peat. This is not hard to believe, for documents of the U.S. Department of Agriculture record numerous accounts of such destruction in swamps and bogs during periods of intense drought.

Other problems relating to the management of Histosols include erosion and bearing capacity. The very light character of the partially decomposed organic matter, which makes this soil ideal for plant root development, also contributes to its susceptibility to wind erosion. An intense storm with gusty winds rapidly can remove the surface organic horizon. Many farm managers try to counter this problem by maintaining trees as a windbreak on the margins of their fields. The problem of wind erosion increases significantly whenever the soil is left bare, after harvest and prior to spring planting.

Histosols have very low capacities to support weight. Soil subsidence and compaction under weight can cause stresses leading to the cracking and rupturing of constructions such as roads and buildings. Therefore, it is frequently necessary to drive concrete pilings deep enough to reach the subsurface mineral strata for building supports, and in the case of roads the entire organic layer may have to be removed and replaced with a substitute such as sand and/or gravel. Similarly, the farm manager must exercise care in using equipment on these soils. The excessive weight of heavy machinery breaks down structure, compacts the soil, and reduces the ability of plant roots to penetrate.

Appendix: Descriptive Soil Profile Symbols

01 Unaltered organic layer.

02 Decomposed organic layer.

A1 Organic rich, mineral horizon at or adjacent to the surface.

A2 Mineral horizon of maximum eluviation.

A3 A horizon transitional between A and B, and dominated by properties characteristic of an overlying A1 or A2.

B1 A horizon transitional between A and B, and dominated by properties characteristic of an underlying B2.

B2 Mineral horizon of maximum illuviation and/or what has the characteristic upon which the B horizon is most clearly based.

B3 A horizon transitional between B and C or R, and dominated by properties characteristic of the overlying B2.

C Weathered parent material.

R Underlying consolidated bedrock.

b A buried soil layer.

ca An accumulation of calcium carbonate.

cn An accumulation of concretions.

cs An accumulation of calcium sulfate (gypsum).

f Frozen ground.

g A waterlogged (gleyed) layer.

h An accumulation of illuvial humus.

ir An accumulation of illuvial iron.

m An indurated layer, or hardpan, due to silication or calcification.

p   A layer disturbed by plowing.

sa   An accumulation of soluble salts.

si   Cementation by siliceous material.

t   An accumulation of illuvial clay.

x   A fragipan.

Soil color designations specify the relative degrees of three measurable variables: *hue, value,* and *chroma.*

> *Hue* is the dominant spectral color and is related to the wavelength of light. It is symbolized with a letter abbreviation of the color of the rainbow (R = red, Y = yellow, YR = yellow-red), preceded by a number from 0 to 10. For each letter range, the hue becomes more yellow and less red as the number increases.
>
> *Value* is a measure of the relative lightness or intensity of color and is dependent upon the total amount of light reflected. It ranges from 0, for absolute black, to 10, for absolute white.
>
> *Chroma* is the relative purity, strength, or saturation of a color. The code numbers range from 0 (neutral gray) to about 20 (pure grays, white, black).

When coding the color notation, the order is hue, value, chroma, with a space between the hue letter and the succeeding value number, and a slash (/) between the two numbers for value and chroma. For example: 10YR 6/4 is a color with a hue = 10YR, value = 6, and chroma = 4. Comparing this code with a *Munsell Color Chart* provides the designated color as light yellowish-brown.

Glossary

This glossary of soil science terms has been assembled from the following sources: U. S. Department of Agriculture, *The 1957 Yearbook of Agriculture: Soil* (Washington, D.C.: U.S. Govt. Ptg. Office, 1957); National Cooperative Soil Survey, *Soil Taxonomy* (Washington, D.C.: U. S. Govt. Ptg. Office, 1970); and Soil Science Society of America, *Glossary of Soil Science Terms* (Madison, Wis.: Soil Science Society of America, 1973).

A horizon.   See Soil horizon.

ABC soil.   A soil with a complete profile, including A, B, and C horizons.

Absorbing complex.   The materials in the soil that hold water and chemical compounds, mainly on their surfaces. They are chiefly the fine mineral matter and organic matter.

AC soil.   A soil having a profile containing only A and C horizons, with no clearly developed B horizon.

Accelerated erosion.   See Erosion (ii).

Acid soil.   Soil with a pH value < 7.0.

Actinomycetes.   A nontaxonomic term applied to a group of organisms with characteristics intermediate between the simple bacteria and the true fungi. Most soil actinomycetes are unicellular microorganisms that produce a slender branched mycelium and sporulate by segmentation of the entire mycelium or, more commonly, by segmentation of special hyphae. Includes many but not all organisms belonging to the order Actinomycetales.

Additive. A material added to fertilizer to improve its chemical or physical condition. An additive to liquid fertilizer might prevent crystals from forming in the liquid at temperatures where crystallization would normally take place.

Adsorb. Removal of a substance in solution to a solid surface or a separate phase; to accumulate on a surface.

Adsorption. The attachment of compounds or ionic parts of salts to a surface or another phase. Nutrients in solution (ions) carrying a positive charge become attached to (adsorbed by) negatively charged soil particles.

Aerate. To impregnate with a gas, usually air.

Aeration, soil. The process by which air in the soil is replaced by air from the atmosphere. In well-aerated soil, the soil air is very similar in composition to the atmosphere above the soil. Poorly aerated soils usually contain a much higher percentage of carbon dioxide and a correspondingly lower percentage of oxygen than the atmosphere above the soil. The rate of aeration depends largely on the volume and continuity of pores within the soil.

Aeration porosity. *See* Air porosity.

Aerobic. (i) Having molecular oxygen as a part of the environment. (ii) Growing only in the presence of molecular oxygen, as aerobic organisms. (iii) Occuring only in the presence of molecular oxygen (said of certain chemical or biochemical processes such as aerobic decomposition).

Aggregate (of soil). Many fine soil particles held in a single mass or cluster, such as a clod, crumb, block, or prism. Many properties of the aggregate differ from those of an equal mass of unaggregated soil.

Agric horizon. A mineral soil horizon in which clay, silt, and humus derived from an overlying cultivated and fertilized layer have accumulated. The wormholes and illuvial clay, silt, and humus, occupy at least 5 percent of the horizon by volume. The illuvial clay and humus occur as horizontal lamellae or fibers, or as coatings on ped surfaces or in wormholes.

Agronomy. A specialization of agriculture concerned with the theory and practice of field crop production and soil management. The scientific management of land.

Air-dry. (i) The state of dryness (of a soil) at equilibrium with the moisture content in the surrounding atmosphere. The actual moisture content will depend upon the relative humidity and the temperature of the surrounding atmosphere. (ii) To allow to reach equilibrium in moisture content with the surrounding atmosphere.

Air porosity. The proportion of the bulk volume of soil that is filled with air at any given time or under a given condition such as a specified moisture tension.

Albic horizon. A mineral soil horizon from which clay and free iron oxides have been removed or in which the oxides have been segregated to the extent that the color of the horizon is determined primarily by the color of the primary sand and silt particles rather than by coatings on these particles.

Albolls. Mollisols that have an albic horizon immediately below the mollic epipedon. These soils have an argillic or natric horizon and mottles, iron-manganese concretions, or both, within the albic, argillic, or natric horizon (a suborder in the USDA soil taxonomy).

Alfisols. Mineral soils that have umbric or ochric epipedons, argillic horizons, and that hold water at less than 15-bars tension during at least 3 months when the soil is warm enough for plants to grow outdoors. Alfisols have a mean annual soil temperature of less than 8°C or a base saturation in the lower part of the argillic horizon of 35 percent or more when measured at pH 8.2 (an order in the USDA soil taxonomy).

Alkaline soil. Any soil having a pH greater than 7.0.

Alkalinity, soil. The degree of intensity of alkalinity in a soil, expressed by a value greater than 7.0 for the soil pH.

Allitic soil. A soil from which silica has been removed, leaving a dominance of aluminum and iron compounds in the clay fraction.

Alluvial soil. (i) A soil developing from recently deposited alluvium and exhibiting essentially no horizon development or modification of the recently deposited materials. (ii) When capitalized the term refers to a great soil group of the azonal order consisting of soils with little or no modification of the recent sediment in which they are forming (indicated by absence of a B horizon).

Alluvium. Sand, mud, and other sediments deposited on land by streams.

Alumino-silicates. Compounds containing aluminum, silicon, and oxygen atoms as main constituents.

Amendment. (i) An alteration of the properties of a soil, and thereby of the soil, by the addition of substances such as lime, gypsum, sawdust, to the soil for the purpose of making it more suitable for the production of plants. (ii) Any such substance used for this purpose. Strictly speaking, fertilizers constitute a special group of soil amendments.

Amino acids. Nitrogen-containing organic compounds, large numbers of which link together in the formation of a protein molecule. Each amino acid molecule contains one or more amino ($-NH_2$) groups and at least one carboxyl ($-COOH$) group. In addition, some amino acids (cystine and methionine) contain sulfur.

Ammonia. A colorless gas composed of one atom of nitrogen and three atoms of hydrogen. Ammonia liquefied under pressure is used as a fertilizer.

Ammonification. The biochemical process whereby ammoniacal nitrogen is released from nitrogen-containing organic compounds.

Ammonium fixation. Adsorption of ammonium ions (NH_4^+) by the mineral fraction of the soil in forms that cannot be replaced by a neutral potassium salt solution (such as 1NKCl).

Anaerobic. (i) The absence of molecular oxygen. (ii) Growing in the absence of molecular oxygen (such as anaerobic bacteria). (iii) Occurring in the absence of molecular oxygen (as a biochemical process).

Andepts. Inceptisols that have formed either in vitric pyroclastic materials, or have low bulk density and large amounts of amorphous materials, or both. Andepts are not saturated with water long enough to limit their use for most crops (a suborder in the USDA soil taxonomy).

Anhydrous. Dry, or without water. Anhydrous ammonia is water free; in contrast to the water solution of ammonia commonly known as household ammonia.

Anion. An ion carrying a negative charge of electricity.

Anion-exchange capacity. The sum total of exchangeable anions that a soil can adsorb. Expressed as milliequivalents per 100 grams of soil (or of other adsorbing material such as clay).

Angular cobbly. See Coarse fragments *and* Cobbly.

Anthropic epipedon. A surface layer of mineral soil that has the same requirements as the mollic epipedon with respect to color, thickness, organic carbon content, consistence, and base saturation, but that has more than 250ppm of P_2O_5 soluble in 1 percent citric acid, or is dry more than 10 months (cumulative) during the period when not irrigated. The anthropic epipedon forms under long continued cultivation and fertilization.

Anthropic soil. A soil produced from a natural soil or other earthly deposit by human work with new characteristics that make it different from the natural soil. Examples include deep, black surface soils resulting from centuries of manuring, and naturally acid soils that have lost their distinguishing features due to many centuries of liming and use for grass.

Antibiosis. Opposed to living. Antibiotics supress some micro-organisms.

Antibiotic. A substance produced by one species of organism that, in low concentrations, will kill or inhibit growth of certain other organisms.

Apatite. A native phosphate of lime. The name is given to the chief mineral of phosphate rock and the inorganic compound of bone.

Aqua ammonia. A water solution of ammonia.

Aqualfs. Alfisols that are saturated with water for periods long enough to limit their use for most crops other than pasture or woodland unless they are artificially drained. Aqualfs have mottles, iron-manganese concretions or gray colors immediately below the A1 or Ap horizons, and gray colors in the argillic horizon (a suborder in the USDA soil taxonomy).

Aquents. Entisols that are saturated with water for periods long enough to limit their use for most crops other than pasture unless they are artificially drained. Aquents have low chromas or distinct mottles within 50cm of the surface, or are saturated with water at all times (a suborder in the USDA soil taxonomy).

Aquepts. Inceptisols that are saturated with water for periods long enough to limit their use for most crops other than pasture or woodland unless they are artificially drained. Aquepts have either a histic or umbric epipedon and gray colors within 20 inches, or an ochric epipedon underlain by a cambic horizon with gray colors, or have sodium saturation of 15 percent or more (a suborder in the USDA soil taxonomy).

Aquic. A mostly reducing soil moisture regime nearly free of dissolved oxygen due to saturation by groundwater or its capillary fringe and occurring at periods when the soil temperature at 20 inches is above 41°F.

Aquifer. A water-bearing formation through which water moves more readily than in adjacent formations of lower permeability.

Aquods. Spodosols that are saturated with water for periods long enough to limit their use for most crops other than pasture unless they are artificially drained. Aquods may have a histic epipedon, an albic horizon that is mottled or contains a duripan, or mottling or gray colors within or immediately below the spodic horizon (a suborder of the USDA soil taxonomy).

Aquolls. Mollisols that are saturated with water for periods long enough to limit their use for most crops other than pasture unless they are artificially drained. Aquolls may have a histic epipedon, a sodium saturation in the upper part of the mollic epipedon of more than 15 percent that decreases with depth or mottles or gray colors within or immediately below the mollic epipedon (a suborder in the USDA soil taxonomy).

Aquox. Oxisols that have continuous plinthite near the surface, or that are saturated with water sometime during the year if not artificially drained. Aquox have either a histic epipedon, or mottles or colors indicative of poor drainage within the oxic horizon or both (a suborder in the USDA soil taxonomy).

Aquults. Ultisols that are saturated with water for periods long enough to limit their use for most crops other than pasture or woodland unless they are artificially drained. Aquults have mottles, iron-manganese concretions or gray colors immediately below the A1 or Ap horizons, and gray colors in the argillic horizon (a suborder in the USDA soil taxonomy).

Arents. Entisols that contain recognizable fragments of pedogenic horizons that have been mixed by mechanical disturbance. Arents are not saturated with water for periods long enough to limit their use for most crops (a suborder in the USDA soil taxonomy).

Argids. Aridisols that have an argillic or a natric horizon (a suborder in the USDA soil taxonomy).

Argillic horizon. A mineral soil horizon that is characterized by the illuvial accumulation of layer-lattice silicate clays. The argillic horizon has a certain minimum thickness depending on the thickness of the solum, a minimum quantity of clay in comparison with an overlying eluvial horizon depending on the clay content of the eluvial horizon, and usually has coatings of oriented clay on the surface of pores or peds or bridging sand grains.

Arid climate. A very dry climate like that of desert or semidesert regions where there is only enough water for widely spaced desert plants. The limits of precipitation vary widely according to temperature, with an upper limit for cool regions of less than $10''$ and for tropical regions of as much as $20''$. (The precipitation-effectiveness index ranges from 0 to about 16.)

Arid region. Areas where the potential water losses by evaporation and transpiration are greater than the amount of water supplied by precipitation. In the United States this area is broadly considered to be the dry parts of the 17 western states.

Aridic. A soil moisture regime that has no moisture available for plants for more than half the cumulative time that the soil temperature at a depth of $20''$ is above $41°F$, and has no period as long as 90 consecutive days when there is moisture for plants while the soil temperature at $20''$ is continuously above $46.4°F$.

Aridisols. Mineral soils that have an aridic moisture regime, an ochric epipedon, and other pedogenic horizons but no oxic horizon (an order in the USDA soil taxonomy).

Artificial manure. *See* Compost. (In European usage may denote commercial fertilizers.)

Ash. The nonvolatile residue resulting from the complete burning of organic matter. It is commonly composed of oxides of such elements as silicon, aluminum, iron, calcium, magnesium, and potassium.

Assimilation. Conversion of substances taken from the outside into living tissue of plants or animals.

Association, soil. *See* Soil association.

Autochthonous flora. That portion of the microflora presumed to subsist on the more resistant soil organic matter and little affected by the addition of fresh organic materials. Contrast with zymogenous flora.

Autotrophic. Capable of utilizing carbon dioxide and/or carbonates as a sole or major source of carbon and of obtaining energy for carbon reduction and biosynthetic processes from radiant energy (photoautotroph) or oxidation of inorganic substances (chemoautotroph).

Auxins. Organic substances that cause lengthening of the stem when applied in low concentrations to shoots of growing plants.

Available nutrient. That portion of any element or compound in the soil that can be readily absorbed and assimilated by growing plants. (*Available* should not be confused with *exchangeable*.)

Available water. The portion of water in a soil that can be readily absorbed by plant roots. Considered by most workers to be that water held in the soil against a pressure of up to approximately 15 bars.

B horizon.   *See* Soil horizon.

Badland.   A land type generally devoid of vegetation and broken by an intricate maze of narrow ravines, sharp crests, and pinnacles resulting from serious erosion of soft geologic materials. Most common in arid or semiarid regions. A miscellaneous land type.

Banding (of fertilizers).   The placement of fertilizers in the soil in continuous narrow ribbons, usually at specific distances from the seeds or plants. The fertilizer bands are covered by the soil but are not mixed with it.

Bar.   A unit of pressure equal to one million dynes per square centimeter.

Base-saturation percentage.   The extent to which the adsorption complex of a soil is saturated with exchangeable cations other than hydrogen. It is expressed as a percentage of the total cation-exchange capacity.

Basin irrigation (or level borders).   The application of irrigation water to level areas that are surrounded by border ridges or levees. Usually irrigation water is applied at rates greater than the water intake rate of the soil. The water may stand on uncropped soils for several days until the soil is well soaked; then any excess may be used on other fields. The water may stand a few hours on fields having a growing crop.

Basin listing.   A method of tillage that creates small basins by damming lister furrows at regular intervals of about 4′ to 20′. This method is a modification of ordinary listing and is carried out approximately on the contour on nearly level or gently sloping soils as a means of encouraging water to enter the soil rather than to run off the surface.

BC soil.   A soil profile with B and C horizons but with little or no A horizon.

Bedding soil.   Arranging the surface of fields by plowing and grading into a series of elevated beds separated by shallow ditches for drainage.

Bedrock.   The solid rock underlying soils and the regolith in depths ranging from zero (where exposed by erosion) to several hundred feet.

Bench terrace.   *See* Terrace.

Biological interchange.   The interchange of elements between organic and inorganic states in a soil or other substrate through the agency of biological activity. It results from biological decomposition of organic compounds and the liberation of inorganic materials (mineralization); and from the utilization of inorganic materials in the synthesis of microbial tissue (immobilization). Both processes commonly proceed continuously in soils.

Biosequence. A sequence of related soils that differ, one from the other, primarily because of differences in kinds and numbers of soil organisms as a soil-forming factor.

Bisect. A profile of plants and soil showing the vertical and lateral distribution of roots and tops in their natural positions.

Black earth. A term used by some as synonymous with "chernozem;" by others (in Australia) to describe self-mulching black clays.

Blocky soil structure. *See* Soil structure.

Blown-out land. Areas from which all or almost all of the soil and soil material has been removed by wind erosion. Usually barren, shallow depressions with a flat or irregular floor consisting of a more resistant layer and/or an accumulation of pebbles, or a wet zone immediately above a water table. Usually unfit for crop production. A miscellaneous land type.

Blowout. A small area of blown-out land.

Bog iron-ore. Impure ferruginous deposits developed in bogs or swamps by the chemical or biochemical oxidation of iron carried in solution.

Bonds. Chemical forces holding atoms together to form molecules.

Boralfs. Alfisols that have formed in cool places. Boralfs have frigid or cryic but not pergelic temperature regimes, and have udic moisture regimes. Boralfs are not saturated with water for periods long enough to limit their use for most crops (a suborder in the USDA soil taxonomy).

Border-strip. *See* Buffer strip.

Border-strip irrigation. *See* Irrigation methods.

Borolls. Mollisols with a mean annual soil temperature of less than 46.4°F and never dry for 60 consecutive days or more within the 3 months following the summer solstice. Borolls do not contain material that has more than 40 percent $CaCO_3$ equivalent unless they have a calcic horizon, and they are not saturated with water for periods long enough to limit their use for most crops (a suborder in the USDA soil taxonomy).

Bottomland. *See* Floodplain.

Breccia. A rock composed of coarse angular fragments cemented together.

Broad-base terrace.   *See* Terrace.

Buffer, buffering.   Substances in the soil that act chemically to resist changes in reaction or pH. The buffering action is due mainly to clay and very fine organic matter. Highly weathered tropical clays are less active buffers than most less weathered silicate clays. Thus, with the same degree of acidity, or pH, more lime is required to neutralize (i) a clayey soil than a sandy soil, (ii) a soil rich in organic matter than one low in organic matter, or (iii) a sandy loam in, as an example, Michigan than a sandy loam in central Alabama.

Buffer compounds, soil.   The clay, organic matter, and compounds such as carbonates and phosphates that enable the soil to resist appreciable changes in pH.

Buffer strips.   Established strips of perennial grass or other erosion-resisting vegetation, usually on the contour in cultivated fields, to reduce runoff and erosion (also referred to as border strips or field border strips).

Bulk density, soil.   The mass of dry soil per unit bulk volume. The bulk volume is determined before drying to constant weight of 220°F.

Bulk specific gravity.   The ratio of the bulk density of a soil to the mass of unit volume of water.

Bulk volume.   The volume, including the solids and the pores, of an arbitrary soil mass.

Buried soil.   Soil covered by an alluvial, loessal, or other deposit, usually to a depth greater than the thickness of the solum.

C horizon.   *See* Soil horizon.

Calcareous soil.   Soil containing sufficient free calcium carbonate or calcium-magnesium carbonate to effervesce visibly when treated with cold 0.1 N hydrochloric acid.

Calcic horizon.   A mineral soil horizon of secondary carbonate enrichment that is more than 15cm thick, has a calcium carbonate equivalent of more than 15 percent, and has at least 5 percent more calcium carbonate equivalent than the underlying C horizon.

Caliche.   (i) A layer near the surface, more or less cemented by secondary carbonates of calcium or magnesium precipitated from the soil solution. It may occur as a soft thin soil horizon, as a hard thick bed just beneath the solum, or as a surface layer exposed by erosion. Not a geologic deposit. (ii) Alluvium cemented with sodium nitrate, chloride, and/or other soluble salts in the nitrate deposits of Chile and Peru.

Cambic horizon. A mineral soil that has a texture of loamy very fine sand or finer, has soil structure rather than rock structure, contains some weatherable minerals, and is characterized by the alteration or removal of mineral material as indicated by mottling or gray colors, stronger chromas or redder hues than in underlying horizons, or the removal of carbonates. The cambic horizon lacks cementation or induration and has too few evidences of illuviation to meet the requirements of the argillic or spodic horizon.

Capillary fringe. A zone just above the water table (zero gauge pressure) that remains almost saturated. (The extent and the degree of definition of the capillary fringe depends upon the size distribution of pores.)

Carbon-nitrogen ratio. The ratio of the weight of organic carbon to the weight of total nitrogen (mineral plus organic forms) in soil or organic material. Often used synonymously with carbon-organic nitrogen ratio when mineral nitrogen levels are low.

Category. Any one of the ranks of the system of soil classification in which soils are grouped on the basis of their characteristics.

Catena. A sequence of soils of about the same age, derived from similar parent material, and occurring under similar climatic conditions, but having different characteristics due to variation in relief and in drainage.

Cation exchange. The interchange between a cation in solution and another cation on the surface of any surface-active material such as clay colloid or organic colloid.

Cation-exchange capacity. (CEC) The sum total of exchangeable cations that a soil can adsorb. Expressed in milliequivalents per 100 grams or per gram of soil (or of other exchangers such as clay).

Cemented. Indurated; having a hard, brittle consistency because the particles are held together by cementing substances such as humus, calcium carbonate, or the oxides of silicon, iron, and aluminum. The hardness and brittleness persist even when wet.

Chroma. The relative purity, strength, or saturation of a color; directly related to the dominance of the determining wavelength of the light and inversely related to grayness; one of the three variables of color.

Chronosequence. A sequence of related soils that differ, one from the other, in certain properties primarily as a result of time as a soil-forming factor.

Class, soil. A group of soils having a definite range in a particular property such as acidity, degree of slope, texture, structure, land-use capability, degree of erosion, or drainage.

Clay. (i) A soil separate consisting of particles less than .00008″ in equivalent diameter. (ii) A textural class.

Clay films. Coatings of clay on the surfaces of soil peds, and mineral grains and in soil pores (also called clay skins, clay flows, illuviation cutans, argillans, or tonhautchen).

Clay loam. A textural class.

Clay mineral. (i) Naturally occurring inorganic crystalline material found in soils and other earthy deposits, the particles being of clay size; that is, less than .000008″ in diameter. (ii) Material as described under (i), but not limited by particle size.

Clayey. Containing large amounts of clay or having properties similar to those of clay.

Claypan. A dense, compact layer in the subsoil having a much higher clay content than the overlying material, from which it is separated by a sharply defined boundary; formed by downward movement of clay or by synthesis of clay in place during soil formation. Claypans are usually hard when dry, and plastic and sticky when wet. Also, they usually impede the movement of water and air, and the growth of plant roots.

Climosequence. A sequence of related soils that differ, one from the other, in certain properties primarily as a result of the effect of climate as a soil-forming factor.

Clinosequence. A group of related soils that differ, one from the other, in certain properties primarily as a result of the effect of the degree of slope on which they were formed.

Clod. A compact, coherent mass of soil ranging in size from .2″ to as much as 8 or 10″; produced artificially, usually by the activity of man by plowing or digging, especially when these operations are performed on soils that are either too wet or too dry for normal tillage operations.

Coarse fragments. Rock or mineral particles greater than .08 in diameter.

Cobbly. Containing appreciable quantities of cobblestones (said of soil and of land). The term *angular cobbly* is used when the fragments are less rounded.

Colloid. A substance in a state of fine subdivision with particles from 0.00001 to 0.0000001 centimeters (1 micron to 1 millimicron).

Concretion. A local concentration of a chemical compound, such as calcium carbonate or iron oxide, in the form of a grain or nodule of varying size, shape, hardness, and color.

Consistency. (i) The resistance of a material to deformation or rupture. (ii) The degree of cohesion or adhesion of the soil mass. Terms used for describing consistency at various soil moisture contents are:

Wet soil. nonsticky, slightly sticky, sticky, very sticky, non-plastic, slightly plastic, plastic, and very plastic.

Moist soil. loose, very friable, friable, firm, very firm, and extremely firm.

Dry soil. loose, soft, slightly hard, hard, very hard, and extremely hard.

Cementation. weakly cemented, strongly cemented, and indurated.

Crotovina. A former animal burrow in one soil horizon that has been filled with organic matter or material from another horizon (also spelled "krotovina").

Crumb. A soft, porous, more or less rounded ped from .04″ to .2″ in diameter.

Crumb structure. A structural condition in which most of the peds are crumbs.

Crust. A surface layer on soils, ranging in thickness from a few millimeters to perhaps as much as an inch, that is much more compact, hard, and brittle when dry than the material immediately beneath it.

Cryic. A soil temperature regime that has mean annual soil temperatures of more than 37°F but less than 46.4°F, more than 41°F difference between mean summer and mean winter soil temperatures at 20″, and cold summer temperatures.

Crystal structure. The orderly arrangement of atoms in a crystalline material.

Crystalline rock. A rock consisting of various minerals that have crystallized in place from magma.

Cultivation. A tillage operation used in preparing land for seeding or transplanting or later for weed control and for loosening the soil.

Decomposition. The process of resolving into constituent parts. Mineral elements are generally among the terminal products.

Deflation. The removal of fine soil particles from soil by wind.

Deflocculate. (i) To separate the individual components of compound particles by chemical and/or physical means. (ii) To cause the particles of the disperse phase of a colloidal system to become suspended in the dispersion medium.

Denitrification. The biochemical reduction of nitrate or nitrite to gaseous nitrogen either as molecular nitrogen or as an oxide of nitrogen.

Deposit. Material left in a new position by a natural transporting agent such as water, wind, ice, or gravity, or by the activity of man.

Desert crust. A hard layer, containing calcium carbonate, gypsum, or other binding material, exposed at the surface in desert regions.

Desert pavement. The layer of gravel or stones left on the land surface in desert regions after the removal of the fine material by wind erosion.

Durinodes. Weakly cemented to indurated soil nodules cemented with SiO_2. Durinodes break down in concentrated KOH after treatment with HCl to remove carbonates, but do not break down on treatment with concentrated HCl alone.

Duripan. A mineral soil horizon that is cemented by silica, usually opal or microcrystalline forms of silica, to the point that air-dry fragments will not slake in water or HCl. A duripan may also have accessory cement such as iron oxide or calcium carbonate.

Dust mulch. A loose, finely granular, or powdery condition on the surface of the soil, usually produced by shallow cultivation.

Edaphic. (i) Of or pertaining to the soil. (ii) Resulting from or influenced by factors inherent in the soil or other substrate, rather than by climatic factors.

Edaphology. The science that deals with the influence of soils on living things, particularly plants, including human use of land for plant growth.

Effective precipitation. The portion of the total precipitation that becomes available for plant growth.

Eluvial horizon. A soil horizon that has been formed by the process of eluviation.

Eluviation. The removal of soil material in suspension (or in solution) from a layer or layers of a soil. (Usually, the loss of material in solution is described by the term *leaching*.)

Entisols. Mineral soils that have no distinct pedogenic horizons within 40″ of the soil surface (an order in the USDA soil taxonomy).

Erode. To wear away or remove the land surface by wind, water, or other agents.

Erodible. Susceptible to erosion (expressed by terms such as highly erodible, slightly erodible, and others).

Erosion. (i) The wearing away of the land surface by running water, wind, ice, or other geological agents, including such processes as gravitational creep. (ii) Detachment and movement of soil or rock by water, wind, ice, or gravity.

Evapotranspiration. The combined loss of water from a given area, and during a specified period of time, by evaporation from the soil surface and by transpiration from plants.

Exchange capacity. The total ionic charge of the adsorption complex active in the adsorption of ions.

Family, soil. In soil classification one of the categories intermediate between the great soil group and the soil series.

Ferrods. Spodosols that have more than six times as much free iron (elemental) than organic carbon in the spodic horizon. Ferrods are rarely saturated with water or do not have characteristics associated with wetness.

Fertility, soil. The status of a soil with respect to the amount and availability to plants of elements necessary for plant growth.

Fertilizer. Any organic or inorganic material of natural or synthetic origin that is added to a soil to supply certain elements essential to the growth of plants.

Fibrists. Histosols that have a high content of undecomposed plant fibers and a bulk density less than about 0.1. Fibrists are saturated with water for periods long enough to limit their use for most crops unless they are artificially drained.

Field border strips. *See* Buffer strips.

Field capacity (field moisture capacity). The percentage of water remaining in a soil 2 or 3 days after having been saturated and after free drainage has practically ceased. (The percentage may be expressed on the basis of weight or volume.)

Film water. A layer of water surrounding soil particles and varying in thickness from 1 or 2 to perhaps 100 or more molecular layers. Usually considered as that water remaining after drainage has removed free water, because it is not distinguishable in saturated soils.

Fine texture. Consisting of or containing large quantities of the fine fractions, particularly of silt and clay.

Firm. A term describing the consistency of a moist soil that offers distinctly noticeable resistance to crushing but can be crushed with moderate pressure between the thumb and forefinger.

First bottom. The normal floodplain of a stream.

Fixation. The process or processes in a soil by which certain chemical elements essential for plant growth are converted from a soluble or exchangeable form to a much less soluble or to a nonexchangeable form; for example, phosphate "fixation." Contrast with nitrogen fixation.

Fluvents. Entisols that form in recent loamy or clayey alluvial deposits, are usually stratified, and have an organic carbon content that decreases irregularly with depth. Fluvents are not saturated with water for periods long enough to limit their use for most crops.

Folists. Histosols that have an accumulation of organic soil materials mainly as forest litter that is less than 40″ deep to rock or to fragmental materials with interstices filled with organic materials. Folists are not saturated with water for periods long enough to limit their use if cropped.

Fragipan. A natural subsurface horizon with high bulk density relative to the solum above, seemingly cemented when dry, but when moist showing a moderate to weak brittleness. The layer is low in organic matter, mottled, slowly or very slowly permeable to water, and usually shows occasional or frequent bleached cracks forming polygons. It may be found in profiles of either cultivated or virgin soils but not in calcareous material.

Friable. A consistency term pertaining to the ease of crumbling of soils.

Frigid. A soil temperature regime that has mean annual soil temperatures of more than 32°F but less than 46.4°F, more than 41°F difference between mean summer and mean winter soil temperatures at 20″, and warm summer temperatures. Isofrigid is the same, except the summer and winter temperatures differ by less than 41°F.

Fulvic acid. A term of varied usage but usually referring to the mixture of organic substances remaining in solution upon acidification of a dilute alkali extract from the soil.

Genetic. Resulting from, or produced by, soil-forming processes; for example, a genetic soil profile or a genetic horizon altered from its geologic form.

Gilgai. The microrelief of soils produced by expansion and contraction with changes in moisture. Found in soils that contain large amounts of clay that swell and shrink considerably with wetting and drying. Usually a succession of microbasins and microknolls in nearly level areas or of microvalleys and microridges parallel to the direction of the slope.

Gravelly. Containing appreciable or significant amounts of gravel (used to describe soils or lands).

Gravitational water. Water which moves into, through, or out of the soil under the influence of gravity.

Green manure. Plant material incorporated with the soil while green, or soon after maturity, for improving the soil.

Green-manure crop. A crop grown for use as green manure.

Groundwater. The portion of the total precipitation that at any particular time is either passing through or standing in the soil and the underlying strata and is free to move under the influence of gravity.

Gypsic horizon. A mineral soil horizon of secondary calcium sulfate enrichment that is more than 6″ thick, has at least 5 percent more gypsum than the C horizon, and in which the product of the thickness in centimeters and the percent calcium sulfate is equal to or greater than 150 percent cm.

Hardpan. A hardened soil layer, in the lower A or in the B horizon, caused by cementation of soil particles with organic matter or with materials such as silica, sesquioxides, or calcium carbonate. The hardness does not change appreciably with changes in moisture content, and pieces of the hard layer do not slake in water.

Hemists. Histosols that have an intermediate degree of plant fiber decomposition and a bulk density between about 0.1 and 0.2. Hemists are saturated with water for periods long enough to limit their use for most crops unless they are artificially drained.

Heterotrophic. An organism capable of deriving energy for life processes from the oxidation of organic compounds.

Histic epipedon. A thin organic soil horizon that is saturated with water at some period of the year unless artificially drained and that is at or near the surface of a mineral soil. The histic epipedon has a maximum thickness depending on the kind of materials in the horizon, and the lower limit of organic carbon is the upper limit for the mollic epipedon.

Histosols. Organic soils that have organic soil materials in more than half of the upper 32″, or that are of any thickness of overlying rock or fragmental materials that have interstices filled with organic soil materials.

Hue.   One of the three variables of color. It is caused by light of certain wavelengths and changes with the wavelength.

Humic acid.   A mixture of dark-colored substances of indefinite composition extracted from soil with dilute alkali and precipitated by acidification.

Humification.   The processes involved in the decomposition of organic matter and leading to the formation of humus.

Humods.   Spodosols that have accumulated organic carbon and aluminum, but not iron, in the upper part of the spodic horizon. Humods are rarely saturated with water or do not have characteristics associated with wetness.

Humox.   Oxisols that are moist all or most of the time and that have a high organic carbon content within the upper meter. Humox have a mean annual soil temperature of less than 71.6°F and a base saturation within the oxic horizon of less than 35 percent, measured at pH 7.

Humults.   Ultisols that have a high content of organic carbon. Humults are not saturated with water for periods long enough to limit their use for most crops.

Humus.   (i) That more or less stable fraction of the soil organic matter remaining after the major portion of added plant and animal residues has decomposed. Usually it is dark colored. (ii) Includes the F and H layers in undisturbed forest soils.

Hydrologic cycle.   The fate of water from the time of precipitation until the water has been returned to the atmosphere by evaporation and is again ready to be precipitated.

Hyperthermic.   A soil temperature regime that has mean annual soil temperatures of 71.6°F or more and more than 41°F difference between mean summer and mean winter soil temperatures at 20″. Isohyperthermic is the same except the summer and winter temperatures differ by less than 41°F.

Igneous rock.   Rock formed from the cooling and solidification of magma, and that has not been changed appreciably since its formation.

Illite.   A hydrous mica.

Illuvial horizon.   A soil layer or horizon in which material carried from an overlying layer has been precipitated from solution or deposited from suspension. The layer of accumulation.

Illuviation. The process of deposition of soil material removed from one horizon to another in the soil; usually from an upper to a lower horizon in the soil profile.

Immature soil. A soil with indistinct or only slightly developed horizons because of the relatively short time it has been subjected to the various soil-forming processes. A soil that has not reached equilibrium with its environment.

Immobilization. The conversion of an element from the inorganic to the organic form in microbial tissues or in plant tissues, thus rendering the element not readily available to other organisms or to plants.

Impeded drainage. A condition that hinders the movement of water through soils under the influence of gravity.

Impervious. Resistant to penetration by fluids or by roots.

Inceptisols. Mineral soils that have one or more pedogenic horizons in which mineral materials other than carbonates or amorphous silica have been altered or removed but not accumulated to a significant degree. Under certain conditions, Inceptisols may have an ochric, umbric, histic, plaggen, or mollic epipedon. Water is available to plants more than half of the year or more than 3 consecutive months during a warm season.

Infiltration. The downward entry of water into the soil.

Infiltration rate. A soil characteristic determining or describing the maximum rate at which water can enter the soil under specified conditions, including the presence of an excess of water. It has the dimensions of velocity.

Intergrade. A soil that possesses moderately well-developed distinguishing characteristics of two or more genetically related great soil groups.

Ions. Atoms, groups of atoms, or compounds, which are electrically charged as a result of the loss of electrons (cations) or the gain of electrons (anions).

Iron pan. An indurated soil horizon in which iron oxide is the principal cementing agent.

Irrigation. The artificial application of water to the soil for the benefit of growing crops.

Kaolin. (i) An aluminosilicate mineral of the 1:1 crystal lattice group; that is, consisting of one silicon tetrahedral layer and one aluminum oxide-hydroxide octahedral layer. (ii) The 1:1 group or family of aluminosilicates.

Lattice.   A three-dimensional grid of lines connecting the points representing the centers of atoms or ions in a crystal.

Leaching.   The removal of materials in solution from the soil.

Lime, agricultural.   A soil amendment consisting principally of calcium carbonate but including magnesium carbonate and perhaps other materials, and used to furnish calcium and magnesium as essential elements for the growth of plants and to neutralize soil acidity.

Lime concretion.   An aggregate of precipitated calcium carbonate, or of other material cemented by precipitated calcium carbonate.

Lime pan.   A hardened layer cemented by calcium carbonate.

Lime requirement.   The mass of agricultural limestone, or the equivalent of other specified liming material, required per acre to a soil depth of 6″ (or on 2 million pounds of soil) to raise the pH of the soil to a desired value under field conditions.

Lithic contact.   A boundary between soil and continuous, coherent, underlying material. The underlying material must be sufficiently coherent to make hand digging with a spade impractical. If mineral, it must have a hardness of 3 or more (Mohs scale), and gravel-size chunks that can be broken out and do not disperse with 15 hours shaking in water or sodium hexametaphosphate solution.

Loam.   A soil textural class.

Loamy.   Intermediate in texture, and properties between fine-textured and coarse-textured soils.

Loess.   Material transported and deposited by wind and consisting of predominantly silt-sized particles.

Magma.   A natural occurring silicate melt, which may contain suspended silicate crystals, or dissolved gases, or both.

Marl.   Soft and unconsolidated calcium carbonate, usually mixed with varying amounts of clay or other impurities.

Mature soil.   A soil with well-developed soil horizons produced by the natural processes of soil formation and essentially in equilibrium with its present environment.

Maximum water-holding capacity.   The average moisture content of a disturbed sample of soil, 2.5″ high, which is at equilibrium with a water table at its lower surface.

Medium texture.   Intermediate between fine-textured and coarse-textured (soils).

Mesic. A soil temperature regime that has mean annual soil temperatures of 46.4°F or more but less than 56°F, and more than 41°F difference between mean summer and mean winter soil temperatures at 20″. Isomesic is the same except the summer and winter temperatures differ by less than 41°F.

Mesophilic bacteria. Bacteria whose optimum temperature for growth falls in an intermediate range of approximately 59° to 113°F.

Metamorphic rock. Rock derived from pre-existing rocks but that differ from them in physical, chemical, and mineralogical properties as a result of natural geological processes, principally heat and pressure, originating within the earth. The pre-existing rocks may have been igneous, sedimentary, or another form of metamorphic rock.

Microclimate. (i) The climatic condition of a small area resulting from the modification of the general climatic conditions by local differences in elevation or exposure. (ii) The sequence of atmospheric changes within a very small region.

Microfauna. Protozoa and smaller nematodes.

Microflora. Bacteria, including actinomycetes, viruses, and fungi.

Micronutrient. A chemical element necessary in only extremely small amounts (less than 1 ppm in the plant) for the growth of plants. Examples are: B, Cl, Cu, Fe, Mn, and Zn. (*Micro* refers to the amount used rather than to its essentiality.)

Microrelief. Small-scale, local differences in topography, including mounds, swales, or pits, that are only a few feet in diameter and with elevation differences of up to 6′.

Mineral soil. A soil consisting predominantly of, and having its properties determined predominantly by, mineral matter. Usually contains less than 20 percent organic matter, but may contain an organic surface layer up to 12″ thick.

Mineralization. The conversion of an element from an organic form to an inorganic state as a result of microbial decomposition.

Mineralogical analysis. The estimation or determination of the kinds or amounts of minerals present in a rock or in a soil.

Mollic epipedon. A surface horizon of mineral soil that is dark colored and relatively thick, contains at least 0.58 percent organic carbon, is not massive and hard or very hard when dry, has a base saturation of more than 50 percent when measured at pH 7, has less than 250 ppm of P_2O_5 soluble in 1 percent citric acid, and is dominately saturated with bivalent cations.

Mollisols.   Mineral soils that have a mollic epipedon overlying mineral material with a base saturation of 50 percent or more when measured at pH 7. Mollisols may have an argillic, natric, albic, cambic, gypsic, calcic, or petrocalcic horizon, a histic epipedon, or a duripan, but not an oxic or spodic horizon.

Montmorillonite.   An aluminosilicate clay mineral with a 2:1 expanding crystal structure that is, with two silicon tetrahedral layers enclosing an aluminum octahedral layer. Considerable expansion may be caused along the C axis by water moving between silica layers of continguous units.

Mottled zone.   A layer that is marked with spots or blotches of different color or shades of color. The pattern of mottling and the size, abundance, and color contrast of the mottles may vary considerably and should be specified in soil description.

Mottling.   Spots or blotches of different color or shades of color interspersed with the dominant color.

Muck.   Highly decomposed organic material in which the original plant parts are not recognizable. Contains more mineral matter and is usually darker in color than peat.

Muck soil.   (i) A soil containing between 20 and 50 percent of organic matter. (ii) An organic soil in which the organic matter is well decomposed.

Mulch.   (i) Any material such as straw, sawdust, leaves, plastic film, loose soil that is spread upon the surface of the soil to protect it and plant roots from the effects of raindrops, soil crusting, freezing, evaporation. (ii) To apply mulch to the soil surface.

Munsell color system.   A color designation system that specifies the relative degrees of the three simple variables of color: hue, value, and chroma. For example: 10YR 6/4 is a color (of soil) with a hue = 10YR, value = 6, and chroma = 4. These notations can be translated into several different systems of color names as desired.

Mycorrhiza.   Literally "fungus root." The association, usually symbiotic, of specific fungi with the roots of higher plants.

Natric horizon.   A mineral soil horizon that satisfies the requirements of an argillic horizon, but that also has prismatic, columnar, or blocky structure and a subhorizon having more than 15 percent saturation with exchangeable sodium.

Neutral soil.   A soil in which the surface layer, at least to normal plow depth, is neither acid nor alkaline in reaction.

Nitrification.   Biological oxidation of ammonium to nitrite and nitrate, or a biologically induced increase in the oxidation state of nitrogen.

Nitrogen fixation. Biological conversion of molecular nitrogen (N_2) to organic combinations or to forms utilizable in biological processes.

Ochrepts. Inceptisols formed in cold or temperate climates and that commonly have an ochric epipedon and a cambic horizon. They may have an umbric or mollic epipedon less than $10''$ thick or a fragipan or duripan under certain conditions. These soils are not dominated by amorphous materials and are not saturated with water for periods long enough to limit their use for most crops.

Ochric epipedon. A surface horizon of mineral soil that is too light in color, too high in chroma, too low in organic carbon, or too thin to be a plaggen, mollic, umbric, anthropic, or histic epipedon, or that is both hard and massive when dry.

Organic soil. A soil that contains a high percentage of organic matter throughout the solum.

Organic soil materials. Soil materials that are saturated with water and have 17.4 percent or more organic carbon if the mineral fraction is 50 percent or more clay, or 11.6 percent organic carbon if the mineral fraction has no clay, or has proportional intermediate contents, or if never saturated with water, has 20.3 percent or more organic carbon.

Orthents. Entisols that have either textures of very fine sand or finer in the fine earth fraction, or textures of loamy fine sand or coarser and a coarse fragment content of 35 percent or more and that have an organic carbon content that decreases regularly with depth. Orthents are not saturated with water for periods long enough to limit their use for most crops.

Orthids. Aridisols that have a cambic, calcic, petrocalcic, gypsic, or salic horizon or a duripan, but that lack an argillic or natric horizon.

Orthods. Spodosols that have less than six times as much free iron (elemental) than organic carbon in the spodic horizon but the ratio of iron to carbon is 0.2 or more. Orthods are not saturated with water for periods long enough to limit their use for most crops.

Orthox. Oxisols that are moist all or most of the time, and that have a low to moderate content of organic carbon within the upper $40''$ or a mean annual soil temperature of $71.6°F$ or more.

Ortstein. An indurated layer in the B horizon of podzols in which the cementing material consists of illuviated sesquioxides (mostly iron) and organic matter.

Oxic horizon. A mineral soil horizon that is at least 12″ thick and characterized by the virtual absence of weatherable primary minerals or 2:1 base lattice clays, the presence of 1:1 lattice clays, and highly insoluble minerals such as quartz sand, the presence of hydrated oxides of iron and aluminum, the absence of water-dispersible clay, and the presence of low cation-exchange capacity and small amounts of exchangeable bases.

Oxisols. Mineral soils that have an oxic horizon within 80″ of the surface or plinthite as a continuous phase within 12″ of the surface, and that do not have a spodic or argillic horizon above the oxic horizon.

Paleosol, buried. A soil formed on a landscape during the geologic past and subsequently buried by sedimentation.

Paleosol, exhumed. A formerly buried paleosol that has been exposed on the landscape by the erosive stripping of an overlying mantle of sediment.

Pan, genetic. A natural subsurface soil layer of low or very low permeability, with a high concentration of small particles, and differing in certain physical and chemical properties from the soil immediately above or below the pan.

Pan, pressure or induced. A subsurface horizon or soil layer having a higher bulk density and a lower total porosity than the soil directly above or below it, as a result of pressure that has been applied by normal tillage operations or by other artificial means. Frequently referred to as plowpan, plow sole, or traffic pan.

Pans. Horizons or layers in soils that are strongly compacted, indurated, or very high in clay content.

Paralithic contact. Similar to a lithic contact except that the mineral material below the contact has a hardness of less than 3 (Mohs scale), and gravel-size chunks that can be broken out will partially disperse within 15 hours shaking in water or sodium hexametaphosphate solution.

Parent material. The unconsolidated and more or less chemically weathered mineral or organic matter from which the solum of soils is developed by pedogenic processes.

Particle density. The mass per unit volume of the soil particles. In technical work, usually expressed as grams per cubic centimeter.

Particle size.   The effective diameter of a particle measured by sedimentation, sieving, micrometric methods.

Particle-size analysis.   Determination of the various amounts of the different separates in a soil sample, usually by sedimentation, sieving, micrometry, or combination of these methods.

Peat.   Unconsolidated soil material consisting largely of undecomposed, or only slightly decomposed, organic matter accumulated under conditions of excessive moisture.

Peat soil.   An organic soil containing more than 50 percent organic matter. Used in the United States to refer to the stage of decomposition of the organic matter; "peat" referring to the slightly decomposed or undecomposed deposits, and "muck" to the highly decomposed materials.

Ped.   A unit of soil structure, such as an aggregate, crumb, prism, block, or granule, formed by natural processes (in contrast with a clod, which is formed artificially).

Pedogenic.   *See* Genetic.

Pedon.   A three-dimensional body of soil with lateral dimensions large enough to permit the study of horizon shapes and relations. Its area ranges from 1 to 10 square yards. Where horizons are intermittent or cyclic, and recur at linear intervals of 2 to 7 yards, the pedon includes one-half of the cycle. Where the cycle is less than 2 yards, or all horizons are continuous and of uniform thickness, the pedon has an area of approximately 1 square yard. If the horizons are cyclic, but recur at intervals greater than 7 yards, the pedon reverts to the 1-square-yard size, and more than one soil will usually be represented in each cycle.

Percolation, soil water.   The downward movement of water through soil. Especially, the downward flow of water in saturated or nearly saturated soil at hydraulic gradients of the order of 1.0 or less.

Pergelic.   A soil temperature regime that has mean annual soil temperatures of less than 32°F. Permafrost is present.

Permafrost.   (i) Permanently frozen material underlying the solum. (ii) A perennially frozen soil horizon.

Permeability, soil.   (i) The ease with which gases, liquids, or plant roots penetrate or pass through a bulk mass of soil or a layer of soil. Since different soil horizons vary in permeability, the particular horizon under question should be designated. (ii) The property of a porous medium itself that relates to the ease with which gases, liquids, or other substances can pass through it.

Petrocalcic horizon.   A continuous, indurated calcic horizon that is cemented by calcium carbonate and, in some places, with magnesium carbonate. It cannot be penetrated with a spade or auger when dry; dry fragments do not slake in water; and it is impenetrable to roots.

Petrogypsic horizon. A continuous, strongly cemented, massive, gypsic horizon that is cemented by calcium sulfate. It can be chipped with a spade when dry. Dry fragments do not slake in water and it is impenetrable to roots.

pH, soil. The negative logarithm of the hydrogen-ion activity of a soil. The degree of acidity (or alkalinity) of a soil as determined by means of a glass, quinhydrone, or other suitable electrode, or indicator at a specified moisture content or soil-water ratio, and expressed in terms of the pH scale.

Phase, soil. A subdivision of a soil type or other unit of classification having characteristics that affect the use and management of the soil but which do not vary sufficiently to differentiate it as a separate type. A variation in a property or characteristic, such as degree of slope, degree of erosion, content of stones.

Physical weathering. The breakdown of rock and mineral particles into smaller particles by physical forces such as frost action.

Placic horizon. A black to dark reddish mineral soil horizon that is usually thin but that may range from .04″ to 1″ in thickness. The placic horizon is commonly cemented with iron and is slowly permeable or impenetrable to water and roots.

Plaggen epipedon. A man-made surface horizon more than 20″ thick that is formed by long-continued manuring and mixing.

Plaggepts. Inceptisols that have a plaggen epipedon.

Plastic soil. A soil capable of being molded or deformed continuously and permanently by relatively moderate pressure into various shapes.

Platy. Consisting of soil aggregates that are developed predominately along the horizontal axis; laminated; flaky.

Plinthite. A nonindurated mixture of iron and aluminum oxides, clay, quartz, and other diluents that commonly occurs as red soil mottles usually arranged in platy, polygonal, or reticulate patterns. Plinthite changes irreversibly to ironstone hardpans or irregular aggregates on exposure to repeated wetting and drying.

Pore-size distribution. The volume of the various sizes of pores in a soil. Expressed as percentages of the bulk volume (soil plus pore space).

Pore space. Total space not occupied by soil particles in a bulk volume of soil.

Porosity. The volume percentage of the total bulk not occupied by solid particles.

Precipitation interception. The stopping, interrupting, and temporary holding of precipitation in any form by a vegetative canopy or vegetation residue.

Primary mineral. A mineral that has not been altered chemically since deposition and crystallization from molten lava.

Prismatic soil structure. A soil structure type with prismlike aggregates that have a vertical axis much longer than the horizontal axes.

Productive soil. A soil in which the chemical, physical, and biological conditions are favorable for the economic production of crops suited to a particular area.

Productivity, soil. The capacity of a soil, in its normal environment, for producing a specified plant or sequence of plants under a specified system of management. The "specified" limitations are necessary since no soil can produce all crops with equal success nor can a single system of management produce the same effect on all soils. Productivity emphasizes the capacity of soil to produce crops and should be expressed in terms of yields.

Profile, soil. A vertical section of the soil through all its horizons and extending into the parent material.

Psamments. Entisols that have textures of loamy fine sand or coarser in all parts, have less than 35 percent coarse fragments, and are not saturated with water for periods long enough to limit their use for most crops.

Reaction, soil. The degree of acidity or alkalinity of a soil, usually expressed as a pH value. Descriptive terms commonly associated with certain ranges in pH are: extremely acid, less than 4.5; very strongly acid, 4.5-5.0; strongly acid, 5.1-5.5; medium acid, 5.6-6.0; slightly acid, 6.1-6.5; neutral, 6.6-7.3; slightly alkaline, 7.4-7.8; moderately alkaline, 7.9-8.4; strongly alkaline, 8.5-9.0; and very strongly alkaline, greater than 9.1.

Regolith. The unconsolidated mantle of weathered rock and soil material on the earth's surface; loose earth materials above solid rock. (Approximately equivalent to the term "soil" as used by many engineers.)

Rendolls. Mollisols that have no argillic or calcic horizon but that contain material with more than 40 percent $CaCO_3$ equivalent within or immediately below the mollic epipedon. Rendolls are not saturated with water for periods long enough to limit their use for most crops.

Residual material.   Unconsolidated and partly weathered mineral materials accumulated by disintegration of consolidated rock in place.

Residual soil.   A soil formed from, or resting on, consolidated rock of the same kind as that from which it was formed, and in the same location. It is also referred to as sedentary soil.

Reticulate mottling.   A network of streaks of different color, most commonly found in the deeper profiles of Lateritic soils.

Rhizobia.   Bacteria capable of living symbiotically in roots of legumes, from which they receive energy and often utilize molecular nitrogen. Collective common name for the gensus *Rhizobium*.

Rhizosphere.   The zone of soil where the microbial population is altered both quantitatively and qualitatively by the presence of plant roots.

Rill.   A small, intermittent water course with steep sides; usually only a few inches deep and, hence, no obstacle to tillage operations.

Runoff.   That portion of the precipitation on an area discharged from the area through stream channels. That which is lost without entering the soil is called surface runoff, and that which enters the soil before reaching the stream is called groundwater runoff or seepage flow from groundwater. (In soil science runoff usually refers to the water lost by surface flow; in geology and hydraulics runoff usually includes both surface and subsurface flow.)

Salic horizon.   A mineral soil horizon of enrichment with secondary salts more soluble in cold water than gypsum. A salic horizon is $3''$ or more in thickness, contains at least 2 percent salt, and the product of the thickness in centimeters and percent salt by weight is 60 percent cm or more.

Sand.   (i) A soil particle between 0.002 and $.08''$ in diameter. (ii) Any one of five soil separates; namely, very coarse sand, coarse sand, medium sand, fine sand, and very fine sand. (iii) A soil textural class.

Sandy.   Containing a large amount of sand. (Applied to any one of the soil classes that contains a large percentage of sand.)

Saprists.   Histosols that have a high content of plant materials so decomposed that original plant structures cannot be determined, and a bulk density of about 0.2 or more. Saprists are saturated with water for periods long enough to limit their use for most crops unless they are artificially drained.

Saturate.   (i) To fill all the voids between soil particles with a liquid. (ii) To form the most concentrated solution possible under a given set of physical conditions in the presence of an excess of the solute. (iii) To fill to capacity, as the adsorption complex with a cation species; for example, H-saturated.

Secondary mineral.   A mineral resulting from the decomposition of a primary mineral or from the reprecipitation of the products of decomposition of a primary mineral.

Sedentary soil.   *See* Residual soil.

Sedimentary rock.   A rock formed from materials deposited from suspension or precipitated from solution and usually being more or less consolidated. The principal sedimentary rocks are sandstones, shales, limestones, and conglomerates.

Self-mulching soil.   A soil in which the surface layer becomes so well aggregated that it does not crust and seal under the impact of rain, but instead serves as a surface mulch upon drying.

Shaly.   (i) Containing a large amount of shale fragments, as a soil. (ii) A soil phase as, for example, shaly phase.

Silica-alumina ratio.   The molecules of silicon dioxide (SiO_2) per molecule of aluminum oxide (Al_2O_3) in clay minerals or in soils.

Silt.   (i) A soil separate consisting of particles between .002 and .00008″ in equivalent diameter. (ii) A soil textural class.

Silt loam.   A soil textural class containing a large amount of silt and small quantities of sand and clay.

Silty clay.   A soil textural class containing a relatively large amount of silt and clay and a small amount of sand.

Silty clay loam.   A soil textural class containing a relatively large amount of silt, a lesser quantity of clay, and a still smaller quantity of sand.

Slickensides.   Polished and grooved surfaces produced by one mass sliding past another. Common in Vertisols.

Soil.   (i) The unconsolidated mineral material on the immediate surface of the earth that serves as a natural medium for the growth of land plants. (ii) The unconsolidated mineral matter on the surface of the earth that has been subjected to and influenced by genetic and environmental factors of: parent material, climate (including moisture and temperature effects), macro- and micro-organisms, and topography, all acting over a period of time and producing a product—soil—that differs from the material from which it is derived in many physical, chemical, biological, and morphological properties and characteristics.

Soil air. The soil atmosphere; the gaseous phase of the soil, being that volume not occupied by solid or liquid.

Soil association. (i) A group of defined and named taxonomic soil units occurring together in an individual and characteristic pattern over a geographic region, comparable to plant associations in many ways. (ii) A mapping unit usually used on general soil maps and sometimes on detailed soil maps in which two or more defined taxonomic units occurring together in a characteristic pattern are mapped as a unit because the scale of the map or the purpose for which it is being made does not require delineation of the individual soils.

Soil complex. A mapping unit used in detailed soil surveys where two or more defined taxonomic units are so intimately intermixed geographically that it is undesirable or impractical, because of the scale being used, to separate them.

Soil conservation. (i) Protection of the soil against physical loss by erosion or against chemical deterioration; that is, excessive loss of fertility by either natural or artificial means. (ii) A combination of all management and land-use methods that safeguard the soil against depletion or deterioration by natural or by man-induced factors. (iii) A division of soil science concerned with soil conservation (i) and (ii).

Soil-formation factors. The variable, usually interrelated natural agencies that are active in and responsible for the formation of soil. The factors are usually grouped into five major categories: parent material, climate, organisms, topography, and time.

Soil genesis. (i) The mode of origin of the soil with special reference to the processes or soil-forming factors responsible for the development of the solum, or true soil, from the unconsolidated parent material. (ii) A division of soil science concerned with soil genesis.

Soil geography. A subspecialization of physical geography concerned with the areal distributions of soil types.

Soil horizon. A layer of soil or soil material approximately parallel to the land surface and differing from adjacent genetically related layers in physical, chemical, and biological properties or characteristics such as color, structure, texture, consistency, kinds, and numbers of organisms present, degree of acidity or alkalinity. It is produced by the interaction of soil-forming factors as opposed to strata, which are geologically produced.

Soil management. (i) The sum total of all tillage operations, cropping practices, fertilizers, lime, and other treatments conducted on or applied to a soil for the production of plants. (ii) A division of soil science concerned with the items listed under (i).

Soil mineral. (i) Any mineral that occurs as a part of or in the soil. (ii) A natural inorganic compound with definite physical, chemical, and crystalline properties (within the limits of isomorphism) that occurs in the soil.

Soil mineralogy. A subspecialization of soil science concerned with the homogeneous inorganic materials found in the earth's crust to the depth of weathering or of sedimentation.

Soil moisture. Water contained in the soil.

Soil organic matter. The organic fraction of the soil; includes plant and animal residues at various stages of decomposition, cells and tissues of soil organisms, and substances synthesized by the soil population. Usually determined on soils which have been sieved through a .08″ sieve.

Soil pores. That part of the bulk volume of soil not occupied by soil particles; interstices; voids.

Soil science. That science dealing with soils as a natural resource on the surface of the earth, including soil formation, classification and mapping, and the physical, chemical, biological, and fertility properties of soils per se; and these properties in relation to their management for crop production.

Soil separates. Mineral particles, less than .08″ in equivalent diameter, ranging between specified size limits. The names and size limits of separates recognized in the United States are: very coarse sand, .08 to .04″; coarse sand, .04 to .02″; medium sand, .02 to .01″; fine sand, .01 to .004″; very fine sand, .004 to .002″; silt, .002″ to .00008″; and clay, less than 0.00008″.

Soil series. The basic unit of soil classification being a subdivision of a family and consisting of soils that are essentially alike in all major profile characteristics except the texture of the A horizon.

Soil solution. The aqueous liquid phase of the soil and its solutes consisting of ions dissociated from the surfaces of the soil particles and of other soluble materials.

Soil structure. The combination or arrangement of primary soil particles into secondary particles, units, or peds. These secondary units may be, but usually are not, arranged in the profile in such a manner as to give a distinctive characteristic pattern. The secondary units are characterized and classified on the basis of size, shape, and degree of distinctness into classes, types, and grades, respectively.

Soil structure classes. A grouping of soil structural units or peds on the basis of size.

Soil structure grades. A grouping or classification of soil structure on the basis of inter- and intra-aggregate adhesion, cohesion, or stability within the profile. Four grades of structure designated from 0 to 3 are recognized as follows:

0. *Structureless.* No observable aggregation or no definite and orderly arrangement of natural lines of weakness. Massive, if coherent; single grain, if not coherent.

1. *Weak.* Poorly formed indistinct peds, barely observable in place.

2. *Moderate.* Well-formed distinct peds, moderately durable and evident, but not distinct in undisturbed soil.

3. *Strong.* Durable peds that are quite evident in undisturbed soil, adhere weakly to one another, withstand displacement, and become separated when the soil is disturbed.

Soil survey. The systematic examination, description, classification, and mapping of soils in an area. Soil surveys are classified according to the kind and intensity of field examination.

Soil texture. The relative proportions of the various soil separates in a soil as described by the classes of soil texture. The textural classes may be modified by the addition of suitable adjectives when coarse fragments are present in substantial amounts; for example, "stony silt loam," or "silt loam stony phase." The sand, loamy sand, and sandy loam are further subdivided on the basis of the proportions of the various sand separates present. The limits of the various classes and subclasses are as follows:

Sand. Soil material that contains 85 percent or more of sand; percentage of silt, plus 1.5 times the percentage of clay, shall not exceed 15.

Coarse sand. 25 percent or more very coarse and coarse sand, and less then 50 percent any other one grade of sand.

Sand. 25 percent or more very coarse, coarse, and medium sand, and less than 50 percent or very fine sand.

Fine sand. 50 percent or more fine sand (or) less than 25 percent very coarse, coarse, and medium sand, and less than 50 percent very fine sand.

Very fine sand. 50 percent or more very fine sand.

Loamy sand. Soil material that contains at the upper limit 85 to 90 percent sand, and the percentage of silt plus 1.5 times the percentage of clay is not less than 70 to 85 percent sand, and the percentage of silt plus twice the percentage of clay does not exceed 30.

Loamy coarse sand. 25 percent or more very coarse and coarse sand, and less than 50 percent any other one grade of sand.

Loamy sand. 25 percent or more very coarse, coarse, and medium sand, and less than 50 percent fine or very fine sand.

Loamy fine sand. 50 percent or more fine sand (or) less than 25 percent very coarse, coarse, and medium sand, and less than 50 percent very fine sand.

Loamy very fine sand. 50 percent or more very fine sand.

Sandy loam. Soil material that contains either 20 percent clay or less, and the percentage of silt plus twice the percentage of clay exceeds 30, and 52 percent or more sand; or less than 7 percent clay, less than 50 percent silt, and between 43 percent and 52 percent sand.

Coarse sandy loam. 25 percent or more very coarse and coarse sand, and less than 50 percent any other one grade of sand.

Sandy loam. 30 percent or more very coarse, coarse, and medium sand, but less than 25 percent very coarse sand, and less than 30 percent very fine or fine sand.

Fine sandy loam. 30 percent or more fine sand and less than 30 percent very fine sand (or) between 15 and 30 percent very coarse, coarse, and medium sand.

Very fine sandy loam. 30 percent or more very fine (or) greater than 40 percent fine and very fine sand, at least half of which is very fine sand, and less than 15 percent very coarse, coarse, and medium sand.

Loam. Soil material that contains 7 to 27 percent clay, 28 to 50 percent silt, and less than 52 percent sand.

Silt loam. Soil material that contains 50 percent or more silt and 12 to 27 percent clay (or) 50 to 80 percent silt and less than 12 percent clay.

Silt. Soil material that contains 80 percent or more silt and less than 12 percent clay.

Sandy clay loam. Soil material that contains 20 to 35 percent clay, less than 28 percent silt, and 45 percent or more sand.

Clay loam. Soil material that contains 27 to 40 percent clay and 20 to 45 percent sand.

Silty clay loam. Soil material that contains 27 to 40 percent clay and less than 20 percent sand.

Sandy clay.   Soil material that contains 35 percent or more clay and 45 percent or more sand.

Silty clay.   Soil material that contains 40 percent or more clay and 40 percent or more silt.

Clay.   Soil material that contains 40 percent or more clay, less than 45 percent sand, and less than 40 percent silt.

Solum (plural: *sola*).   The upper and most weathered part of the soil profile; the A and B horizon.

Sombric horizon.   A subsurface mineral horizon that is darker in color than the overlying horizon but that lacks the properties of a spodic horizon. Common in cool, moist soils of high altitude in tropical regions.

Spodic horizon.   A mineral soil horizon that is characterized by the illuvial accumulation of amorphous materials composed of aluminum and organic carbon with or without iron. The spodic horizon has a certain minimum thickness, and a minimum quantity of extractable carbon plus iron plus aluminum in relation to its content of clay.

Spodosols.   Mineral soils that have a spodic horizon or a placic horizon that overlies a fragipan.

Stones.   Rock fragments greater than 10″ in diameter if rounded, and greater than 15″ along the greater axis if flat.

Stoniness.   The relative proportion of stones in or on the soil (used in classification of soils).

Stony.   Containing sufficient stones to interfere with or to prevent tillage. To be classified as stony, greater than 0.01 percent of the surface of the soil must be covered with stones. Used to modify soil class, as stony clay loam or clay loam, stony phase.

Stratified.   Arranged in or composed of strata or layers.

Strip cropping.   The practice of growing crops that requires different types of tillage, such as row and sod, in alternate strips along contours or across the prevailing direction of wind for the purpose of reducing erosion.

Stubble mulch.   The stubble of crops or crop residues left essentially in place on the land as a surface cover before and during the preparation of the seedbed and at least partly during the growing of a succeeding crop.

Subsoiling.   Breaking of compact subsoils, without inverting them, with a special knifelike instrument (chisel) that is pulled through the soil at depths usually of 12″ to 24″ and at spacings usually of 2′ to 5′.

Subsurface tillage. Tillage with a special sweeplike plow or blade that is drawn beneath the surface at depths of several inches and cuts plant roots and loosens the soil without inverting it or without incorporating the surface cover.

Thermic. A soil temperature regime that has mean annual soil temperatures of 59°F or more but less than 71.6°F, and more than 41.0°F difference between mean summer and mean winter soil temperatures at 20″. Isothermic is the same except the summer and winter temperatures differ by less than 41.°F.

Tight soil. A compact, relatively impervious and tenacious soil (or subsoil) that may or may not be plastic.

Tile drain. Concrete, ceramic or plastic pipe placed at suitable depths and spacings in the soil or subsoil to provide water outlets from the soil.

Till. (i) Unstratified glacial drift deposited directly by the ice, and consisting of clay, sand, gravel, and boulders intermingled in any proportion. (ii) To plow and prepare for seeding; to seed or cultivate the soil.

Tilth. The physical condition of soil as related to its ease of tillage, fitness as a seedbed, and its impedance to seedling emergence and root penetration.

Toposequence. A sequence of related soils that differ, one from the other, primarily because of topography as a soil formation factor.

Topsoil. (i) The layer of soil moved in cultivation. (ii) The A horizon. (iii) The A1 horizon. (iv) Presumably fertile soil material used to topdress road banks, gardens, and lawns.

Torrents. Vertisols of arid regions with wide, deep cracks that remain open throughout the year in most years.

Torric. A soil moisture regime defined like aridic moisture regime, but used in a different category of the soil taxonomy.

Torrox. Oxisols that have a torric soil moisture regime.

Transported soil. Any soil that was formed on unconsolidated sedimentary rocks.

Tropepts. Inceptisols that have a mean annual soil temperature of 46.4°F or more, and less than 41°F difference between mean summer and mean winter temperatures at a depth of 20″ below the surface. Tropepts may have an ochric epipedon and a cambic horizon, or an umbric epipedon, or a mollic epipedon under certain conditions, but no plaggen epipedon, and are not saturated with water for periods long enough to limit their use for most crops.

Udalfs. Alfisols that have a udic soil moisture regime and mesic or warmer soil temperature regimes. Udalfs generally have brownish colors throughout, and are not saturated with water for periods long enough to limit their use for most crops.

Uderts. Vertisols of relatively humid regions that have wide, deep cracks that usually remain open continuously for less than 2 months or intermittently for periods that total less than 3 months.

Udic. A soil moisture regime that is neither dry for as long as 90 cumulative days nor for as long as 60 consecutive days in the 90 days following the summer solstice at periods when the soil temperature at 20″ is above 41°F.

Udolls. Mollisols that have a udic soil moisture regime with mean annual soil temperatures of 46.6°F or more. Udolls have no calcic or gypsic horizon, and are not saturated with water for periods long enough to limit their use for most crops.

Udults. Utisols that have low or moderate amounts of organic carbon, reddish or yellowish argillic horizons, and a udic soil moisture regime. Udults are not saturated with water for periods long enough to limit their use for most crops.

Ultisols. Mineral soils that have an argillic horizon with a base saturation of less than 35 percent. Ultisols have a mean annual soil temperature of 46.4°F or higher.

Umbrepts. Inceptisols formed in cold or temperate climates that commonly have an umbric epipedon, but they may have a mollic or an antropic epipedon 1″ or more thick under certain conditions. These soils are not dominated by amorphous materials and are not saturated with water for periods long enough to limit their use for most crops.

Umbric epipedon. A surface layer of mineral soil that has the same requirements as the mollic epipedon with respect to color, thickness, organic carbon content, consistence, structure, and P_2O_5 content, but that has a base saturation of less than 50 percent when measured at pH 7.

Ustalfs. Alfisols that have an ustic soil moisture regime and mesic or warmer soil temperature regimes. Ustalfs are brownish or reddish throughout and are not saturated with water for periods long enough to limit their use for most crops.

Usterts. Vertisols of temperate or tropical regions with wide, deep cracks that usually remain open for periods that total more than 3 months, but do not remain open continuously throughout the year, and have either a mean annual soil temperature of 71.6°F or more or a mean summer and mean winter soil temperature at 20″ that differ by less than 41°F or have cracks that open and close more than once during the year.

Ustic. A soil moisture regime that is intermediate between the aridic and udic regimes and common in temperate subhumid or semiarid regions, or in tropical and subtropical regions with a monsoon climate. A limited amount of moisture is available for plants but occurs at times when the soil temperature is optimum for plant growth.

Ustolls. Mollisols that have an ustic soil moisture regime and mesic or warmer soil temperature regimes. Ustolls may have a calcic, petrocalcic, or gypsic horizon, and are not saturated with water for periods long enough to limit their use for most crops.

Ustox. Oxisols that have an ustic moisture regime and either hyperthermic or isohyperthermic soil temperature regimes or have less than 20 kg organic carbon in the surface cubic meter.

Ustults. Ultisols that have low or moderate amounts of organic carbon, are brownish or reddish throughout, and have an ustic soil moisture regime.

Value, color. The relative lightness or intensity or color and approximately a function of the square root of the total amount of light. One of the three variables of color.

Vertisols. Mineral soils that have 30 percent or more clay, deep wide cracks when dry, and either gilgai microrelief, intersecting slickensides, or wedgeshaped structural aggregates tilted at an angle from the horizontal.

Water table. The upper surface of groundwater or that level below which the soil is saturated with water; locus of points in soil water at which the hydraulic pressure is equal to atmospheric pressure.

Water table, perched. The water table of a saturated layer of soil that is separated from an underlying saturated layer by an unsaturated layer.

Weathering. All physical and chemical changes produced in rocks, at or near the earth's surface, by atmospheric agents.

Wilting point. Same as "permanent wilting percentage," as defined in standard plant physiology texts.

Windbreak. A planting of trees, shrubs, or other vegetation, usually perpendicular or nearly so to the principal wind direction, to protect soil, crops, homesteads, roads, against the effects of winds, such as wind erosion and the drifting of soil and snow.

Xeralfs. Alfisols that have a xeric soil moisture regime. Xeralfs are brownish or reddish throughout.

Xererts. Vertisols of Mediterranean climates with wide, deep cracks that open and close once each year and usually remain open continuously for more than 2 months. Xererts have a mean annual soil temperature of less than 71.6°F.

Xeric. A soil moisture regime common to Mediterranean climates that have moist cool winters and warm dry summers. A limited amount of moisture is present but does not occur at optimum periods for plant growth. Irrigation or summer fallow is commonly necessary for crop production.

Xerolls. Mollisols that have a xeric soil moisture regime. Xerolls may have a calcic, petrocalcic, or gypsic horizon, or a duripan.

Xerults. Ultisols that have low or moderate amounts of organic carbon, are brownish or reddish throughout, and have a xeric soil moisture regime.

Yield, sustained. A continual annual, or periodic, yield of plants or plant material from an area; implies management practices that will maintain the productive capacity of the land.

Bibliography

Abuajamieh, M.M., "The Structure of the Pantano Beds in the Northern Tucson Basin," unpublished Master's thesis. University of Arizona, 1966.

Alexander, John W., *Economic Geography*. Englewood Cliffs, N.J.: Prentice-Hall, Inc., 1964.

Alexander, L. T., and J. G. Cady, "Genesis and Hardening of Laterite in Soils," *U.S.D.A. Technical Bulletin #1281*. Washington, D.C.: Soil Conservation, 1962.

Allison, Franklin E., "Nitrogen and Soil Fertility," *The 1957 Yearbook of Agriculture: Soil.* pp. 85-94. Washington, D.C.: U. S. Government Printing Office, 1957.

Aubert, G., "Soils with Ferruginous or Ferralitic Crusts of Tropical Regions," *Soil Science*, 95, no. 4 (1963), 235-42.

Baldwin, Mark, Charles E. Kellogg, and James Thorp, "Soil Classification," *The 1938 Yearbook of Agriculture: Soils and Man*, pp. 979-1001. Washington, D.C.: U.S. Government Printing Office, 1938.

Barnes, C. P., "Environment of Natural Grassland," *The 1948 Yearbook of Agriculture: Grass*, pp. 45-49. Washington, D.C.: U.S. Government Printing Office, 1948.

Barry, R.G., and R. J. Chorley, *Atmosphere, Weather, and Climate*. New York: Holt, Rinehart and Winston, Inc., 1970.

Basila, M. R., "Hydrogen Bonding Interaction Between Adsorbate Molecules and Surface Hydroxyl Groups on Silica," *Chemical Physics*, 35 (1961), 1151-58.

Blake, W.P., "The Caliche of Southern Arizona," *American Institute of Mining and Engineering*, 31 (1902), 220-26.

Blank, H,R., and E.W. Tynes, "Formation of Caliche in Situ," *Geological Society of America Bulletin*, 75 (1965), 1387-91.

Breazeale, J.F., and H.V. Smith, "Caliche in Arizona," *University of Arizona Agricultural Experimental Station Bulletin*, 131 (1930), 419-41.

Bretz, J.H., and L. Horberg, "Caliche in Southeastern New Mexico," *Journal of Geology*, 57 (1949), 491-511.

Brown, C.H., "The Origin of Caliche on the Northeastern Llano Estacado, Texas," *Journal of Geology*, 64 (1956), 1-15.

Buckman, Harry O., and Nyle C. Brady, *The Nature and Properties of Soils*, New York: Macmillan Co., 1971.

Bunting, B.T., *The Geography of Soil*. Chicago: Aldine Publishing Co., 1967.

Buol, S.W., *Soils of Arizona*. Tucson: Agricultural Experiment Station, University of Arizona, 1966.

_____ "Present Soil-forming Factors and Processes in Arid and Semiarid Regions," Technical Paper no. 895. Tucson: Arizona Agriculture Experiment Station.

Buol, S.W., F.D. Hole, and R.J. McCracken, *Soil Genesis and Classification*. Ames, Iowa: Iowa State University Press, 1973.

Chang, J., "The Agricultural Potential of the Humid Tropics," *The Geographical Review*, LVIII, no. 3 (1968), 333-61.

Clark, Francis E., "Living Organisms in the Soil," *The 1957 Yearbook of Agriculture: Soil*, pp. 157-60. Washington, D.C.: U.S. Government Printing Office, 1957.

Colwell, W.E., "Tobacco," *The 1957 Yearbook of Agriculture: Soil*, pp. 655-58. Washington, D.C.: U.S. Government Printing Office, 1957.

Cooley, D.B., "Geological Environment and Engineering Properties of Caliche in the Tucson Area," unpublished Master's thesis. University of Arizona, 1966.

Cooley, M.E., "Description and Origin of Caliche in the Glen-San Juan Canyon Region, Utah and Arizona," *Arizona Geological Society Digest*, 4 (1961), 35-41.

Donahue, R. L., J. C. Shickluna, and L. S. Robertson, *Soils: An Introduction To Soils and Plant Growth* (3rd ed.). Englewood Cliffs, N.J.: Prentice-Hall, Inc., 1971.

Eeckman, J.P., and H. Laudelett, "Chemical Stability of Hydrogen Montmorilonite," *Kollodzschr*, 178, no. 2 (1961), 99-107.

Eyre, S.R., *Vegetation and Soils*. Chicago: Aldine Publishing Company, 1971.

Foth H.D., and L.M. Turk, *Fundamentals of Soil Science*, 5th ed. New York: John Wiley and Sons, Inc., 1972.

Frederickson, A.F., "Mechanism of Weathering," *Bulletin of the Geologic Society of America*, 62 (1951), 221-32.

Fripiat, J.J., and A.J. Herbillion, "Formation and Transformation of Clay Minerals in Tropical Soils," *Soils and Tropical Weathering*, pp. 15-22. Paris: UNESCO, 1971.

Fuller, W.H., and H.E. Ray, "Gypsum and Sulfur-bearing Amendments," *Agricultural Experiment Station Bulletin*, A-27, pp. 3-11. Tucson: University of Arizona, 1963.

Gibbings, P.N., "The Effect of Caliche on the Strength of Concrete," unpublished Master's thesis. University of Arizona, 1932.

Gile, L.H., "A Classification of ca Horizons in Soils of a Desert Region, Dona Ana County, New Mexico, *Soil Science Society of America Proceedings*, 25 (1961), 52-61.

Gile, L.H., F.F. Peterson, and R.B. Grossman, "The K-horizon: A Master Soil Horizon of Carbonate Accumulation," *Soil Science*, 99 (1965), 74-82.

———, "Morphological and Genetic Sequences of Carbonate Accumulation in Desert Soils," *Soil Science*, 101, (1966), 347-60.

Greenland, D.J., and P.H. Nye, *The Soil Under Shifting Cultivation*. Bucks, England: Commonwealth Agricultural Bureau, 1965.

Grissom, Perrin H., "The Mississippi: Delta Region," *The 1957 Yearbook of Agriculture: Soil*, pp. 524-30. Washington, D.C.: U.S. Government Printing Office, 1957.

Hallsworth, E.G., and A.B. Costin, "Studies in Pedogenesis in New South Wales, IV, 'The Ironstone Soils'," *Soil Science*, 4 (1953), 24-47.

Hendricks, D.M., and Y.H. Havens, *Desert Soils Tour Guide*. Tucson: Soil Science Society of America, 1970.

Hole, Francis D., and Gerald A. Nielson, "Soil Genesis Under Prairie," *Proceedings of the Symposium On Prairie and Prairie Restoration*. Galesburg, Ill.: Knox College, 1963.

Hunt, Charles B. *Geology of Soils: Their Evolution, Classification, and Uses*. San Francisco: W.H. Freeman and Company, 1972.

Isnail, F.T., "Biotite Weathering and Clay Formation in Arid and Humid Regions, California," *Soil Science*, 109 (1969), 257-61.

James, Preston E., *A Geography of Man* (2nd ed.). Boston: Ginn and Company, 1959.

Jenny, H., and C.D. Leonard, "Functional Relationships Between Soil Properties and Rainfall," *Soil Science*, 38 (1934), 363-81.

Keller, W.D., "Process of Origin and Alteration of Clay Minerals," *Soil Clay Minerology*, eds. C.I. Rich and G.W. Kunze. Chapel Hill: University of North Carolina Press, 1964.

Kellogg, Charles E., "Why a New System of Soil Classification?" *Soil Science*, LXXXXCI, no. 1 (July 1963), 1-5.

Martin, W.P., and Joel E. Fletcher, "Vertical Zonation of Great Soil Groups on Mt. Graham, Arizona, as Correlated with Climate, Vegetation, and Profile Characteristics," *Technical Bulletin* 99, pp. 144-47. Tucson: College of Agriculture, University of Arizona, 1943.

Mohrand, E.C.J., and F.A. Van Baren, *Tropical Soils*. London: Interscience Publishers, Ltd., 1954.

Muckernhirn, R.J., and K.C. Berger, "The Northern Lake States," *The 1957 Yearbook of Agriculture: Soil*, pp. 547-53. Washington, D.C.: U.S. Government Printing Office, 1957.

National Cooperative Soil Survey, *Soil Taxonomy*. Washington, D.C.: U.S. Government Printing Office, 1970.

Nikiforoff, C.C., "General Trends of Desert Type of Soil Formation," *Soil Science*, 43, 105-31.

Nye, P.H., "Some Soil-Forming Processes in the Humid Tropics, I-IV," *Journal of Soil Science*, 5, no. 1 (1954), 7-83.

Pearson, R.W., and L.E. Ensminger, "Southeastern Uplands," *The 1957 Yearbook of Agriculture: Soil*, pp. 580-95. Washington, D.C.: U.S. Government Printing Office, 1957.

Pierre, W.H., and F.F. Riecken, "The Midland Feed Region," *The 1957 Yearbook of Agriculture: Soil*, pp. 535-47. Washington, D.C.: U.S. Government Printing Office, 1957.

Post, J.L., "Strength Characteristics of Caliche Soils of the Tucson Area," unpublished Master's thesis. University of Arizona, 1966.

Prescott, J.A., "The Early Use of the Term Laterite," *Journal of Soil Science*, 5, no. 1 (1954), 1-5.

Prescott, J.A., and R.L. Pendleton, *Laterite and Lateritic Soils*. Bucks, England: Commonwealth Agricultural Bureau, 1952.

Price, W.A., "Reynosa Problem of South Texas, and Origin of Caliche," *American Association Petroleum Geologists Bulletin*, 17 (1933), 488-522.

Rao, T. Shesagiri, "Pedogenesis of Some Major Soil Groups in Mysore State, India," *Soils and Tropical Weathering: Proceedings of the Bandung Symposium*, pp. 79-84. Paris: UNESCO, 1971.

Reeves, C.C., Jr., "Caliche," *Encyclopedia of Earth Sciences*, vol. 6. New York: Holt, Rinehart and Winston, 1970.

——, "Origin, Classification, and Geologic History of Caliche on the Southern High Plains, Texas and Eastern New Mexico," *Journal of Geology*, 78 (1970), 352-62.

Reinfenberg, A., and S. J. Buckwold, "The Release of Silica From Soils by the Orthophosphate Anion," *Journal of Soil Science*, 5, no. 1 (1954), 106-15.

Rojan, S.V. Govinda, and N.R. Datta Biswas, "Development of Certain Soils in the Subtropical Humid Zone in Southeastern Parts of India, Genesis and Classification of Soils of Machkund Basin," *Soils and Tropical Weathering*, pp. 77-85. Paris: UNESCO, 1971.

Russell, E.J., *The World of Soil*. London: Collins Clear-Type Press, 1957.

Schuylenborgh, J. Von, "Investigations on the Classification and Genesis of Soils Derived from Andesite Tuffs Under Humid Tropical Conditions," *Netherlands Journal of Agricultural Science*, 5 (1957), 195-210.

Segalen, P., "Metallic Oxides and Hydroxides in Soils of the Warm and Humid Areas of the World: Formation, Identification, Evolution," *Soils and Tropical Weathering*, pp. 25-37. Paris, UNESCO, 1971.

Shreve, F., and T.D. Mallery, "The Relation of Caliche to Desert Plants," *Soil Science*, 35 (1933) 99-113.

Sigalove, J.J., "Carbon 14 Content and Origin of Caliche," unpublished Master's thesis. University of Arizona, 1969.

Simonson, R.W., "Morphology and Classification of the Regur Soils of India," *Journal of Soil Science*, 5, no. 1 (1954), 275-88.

Smith, G.E., "The Physiography of Arizona Valleys and Occurrence of Ground Water," *University of Arizona Agricultural Experiment Station Technical Bulletin*, vol. 77 (1938).

Smith, Guy D., "Lectures on Soil Classification," *Pedologie*, vol. 4. Ghent, Belgium: State University of Ghent, 1965.

Soil Science Society of America, *Glossary of Soil Science Terms*. Madison, Wis.: Soil Science Society of America, 1973.

Soil Survey Staff, *Supplement to Soil Classification, A Comprehensive System, 7th Approximation.* Washington, D.C.: Soil Conservation Service, U.S. Department of Agriculture, 1964.

Stobbe, P.C., and J.R. Wright, "Modern Concepts of the Genesis of Podzols," *Soil Science Society of America Proceedings,* 23 (1959), 161-63.

Stuart, D.M., M.A. Fosburg, and G.C. Lewis, "Caliche in Southwestern Idaho," *Soil Science Society of America Proceedings,* 25 (1961), 132-35.

Syliss, E., G.K. Rennie, C. Smart, and B.A. Pethica, "Anomalous Water," *Nature,* 222 (1969), 159-61.

Thompson, Louis M., *Soils and Soil Fertility.* New York: McGraw-Hill Book Co., 1952.

Thorne, D.W., and L.F. Seaty, "Acid, Alkaline, Alkali, and Saline Soil," *Chemistry of the Soil,* ed. F.E. Bear. New York: Reinhold Publishing Corp., 1955.

Thorne, W., "The Grazing-Irrigated Region," *The 1957 Yearbook of Agriculture,* pp. 481-94. Washington, D.C.: U.S. Government Printing Office, 1957.

Thorp, James, "Effects of Certain Animals That Live in Soils," *Selected Papers In Soil Formation and Classification.* Madison, Wis.: Soils Science Society of America, 1967.

——, "How Soil Develops Under Grass," *The 1948 Yearbook of Agriculture: Grass.* Washington, D.C.: U.S. Government Printing Office, 1948.

Wadia, D.N., "The Pleistocene System," *The Geology of India,* pp. 398-402. London: Macmillan and Co., Ltd., 1953.

Wasser, C.H., Lincoln Ellison, and R.E. Wagner, "Soil Management on Ranges," *The 1957 Yearbook of Agriculture,* pp. 633-41. Washington, D.C.: U.S. Government Printing Office, 1957.

Weaver, John E., and F.E. Clements, *Plant Ecology,* New York: McGraw-Hill Book Company, 1938.

Index